CIRCLING BACK HOME

CIRCLING BACK HOME

A PLAINSWOMAN'S JOURNEY

DARCY LIPP-ACORD

FOREWORD BY LINDA M. HASSELSTROM

SOUTH DAKOTA
STATE HISTORICAL SOCIETY
PRESS / PIERRE

This publication is funded, in part, by the
Great Plains Education Foundation, Inc., Aberdeen, S.Dak.

Library of Congress Cataloging-in-Publication data
Lipp-Acord, Darcy.
Circling back home : a plainswoman's journey / by Darcy Lipp-Acord ;
foreword by Linda M. Hasselstrom.
    pages cm
ISBN 978-0-9860355-5-5 (alk. paper)
1. Lipp-Acord, Darcy. 2. Women ranchers—Wyoming—Biography. 3.
Ranch life—Wyoming. 4. Ranch life—South Dakota. 5. Country life—
Wyoming. 6. Country life—South Dakota. 7. Wyoming—Social life and
customs. 8. South Dakota—Social life and customs. 9. Wyoming—
Biography. 10. South Dakota—Biography. I. Hasselstrom, Linda M. II. Title.
F765.22.L57 2013
978.3'033092—dc23
[B]
        2013021771

The paper in this book meets the guidelines for permanence
and durability of the Committee on Production Guidelines for
Book Longevity of the Council on Library Resources.

Text and cover design by Rich Hendel
Please visit our web site at www.sdshspress.com.

Printed in United States of America
16 15 14 13   1 2 3 4 5

For my family—my foundation;
For my children—my future;
For Shawn—my forever.

CONTENTS

FOREWORD

I know now, in a way I didn't as a teenager, that I belong in the country, the same way my grandmothers did.
— *Darcy Lipp-Acord*

As you drive a lonesome highway chatting on your cell phone, look to the side. See that cowboy riding through the sagebrush? Behind that real cowboy is a real ranch woman like Darcy Lipp-Acord, with real children growing up as the man and his wife juggle real lives.

Cowboys are still part of daily life in South Dakota and other western states. Despite sad songs and frequent news features consigning the Wild West and its inhabitants to the sunset, cowboys ride the pastures surrounding our towns; they haul their children to town school in muddy pickup trucks; they shop in town stores.

The folks who really "belong in the country," as Lipp-Acord puts it, are those who work as well as live there. Unlike most Americans, farmers and ranchers work, live, and usually play at home. What outsiders know about these lives is often wrong. Dime novels, movies, and television have created romantic landscapes: sprawling spreads, ranchers with political power, taciturn cowboys, pretty schoolmarms. Oh yes, and American Indians, usually portrayed as the "Savage Red Man."

Set against these myths are images created by people like Hamlin Garland, who sold his claim in Dakota Territory as quickly as possible to move to Boston and write about the hardships of homesteading. South Dakotans can be grateful to Garland for helping to establish the Friends of the Middle Border (now Dakota Discovery) Museum at Dakota Wesleyan University in Mitchell, which preserves manuscripts from Ole E. Rølvaag, John G. Neihardt, Rose Wilder Lane, L. Frank Baum, and other writers with ties to the region. But Garland's horrendous portrait of prairie life as being filled with endless blizzards, poverty, and hard labor still colors the visions outsiders have of the Dakotas. Either ornamented or brutalized, these myths are not reality.

> *The blue flax dances between life and death,*
> *between holding on and letting go.*
> — *Darcy Lipp-Acord*

Many Americans can count at least one farmer among their ancestors within the last two or three generations. Their lives of labor rarely offered them time, energy, or the connections to publish books about their work, so we have seldom glimpsed the truth about farm and ranch life of the past, especially as lived by women. Relatives of Elizabeth Corey, also known as "Bachelor Bess," saved and published her letters about homesteading. Many women may have scribbled in diaries or written revealing letters, but those documents have been lost or revised to present a gentler and happier view of roughshod reality. Contemporary editors such as Catherine Lavender, Lillian Schlissell, Susan Armitage, Elizabeth Jameson, and others have delved into dusty archives and collected oral histories of ranch and farm women, usually from states more populous than South Dakota. Still, we know little of the viewpoints of most of the hard-working women who settled the prairie with their menfolk. An online term search of "books ranch women" leads directly to Alexa Albert's bitterly ironic commentary on the women of Mustang Ranch, a modern-day Nevada brothel.

Readers may also have formed their opinions from the complaints of the disenchanted, those who failed at farming or ranching on the arid high plains, or from modern writers like Mary Clearman Blew and Judy Blunt, who fully understand the difficulties and prefer another way of life. Some writers who have never lived and certainly never worked on the plains write for profit without regard for truth. By contrast, news media regularly provide accounts of depressing aspects of agriculture: workers who are brutal to their families, who are injured or killed in horrible ways, who commit suicide, who succumb to addiction. We rarely know the continuing stories or hear the voices of the wives and children who are left to grapple with the debts and the sorrow.

The irony sometimes shocks me: after quitting
my teaching career to raise my children, I have become
"just a housewife," like my grandmothers.
—*Darcy Lipp-Acord*

Darcy Lipp-Acord's exploration of her life in the context of her family history helps to set the record straight. A modern ranching woman married to a cowboy, she is the legitimate descendant of authors and ranch women Elinore Pruitt Stewart, Mollie Dorsey Sanford, and countless others who shall be forever unknown. She has made time in her life of labor to write about its details; more importantly, she reflects wisely on what her life means.

Lipp-Acord admits that in college she denied knowing how to cook because she thought domestic skills were not appropriate to a career woman; she even denied having been involved in 4-H, the farm and ranch youth organization. Similarly, many modern career women equate housewifely chores with lack of intelligence or drive, belittling again, ironically, the women who helped to liberate us from these attitudes. How much of our history have we missed by assuming housewives had nothing to teach us?

Rules for life were simple: Work hard.
Don't spend what you don't have. Don't swing
your Rosary beads. Marry another Catholic.
— *Darcy Lipp-Acord*

In various ways, Darcy Lipp-Acord shows us what it means to belong to country life. She grew up in a settled community surrounded by grandparents, uncles and aunts, cousins, and people who had known each other for several generations. Like many of us, she found the closeness confining and escaped to college, planning a life of international travel. And then she fell in love with Shawn, a cowboy who works for wages on large ranches whose owners generally live elsewhere and often regard the cattle business as just an investment. Like his father, Shawn is a professional cowboy, caring for cattle that graze thousands of acres of range. Often, the wives and families of these cowboys are expected to help with ranch work for little or no pay and no benefits such as health or life insurance. The ranch owner may have no idea how to manage a ranch, so the cowboys may be ordered to do unnecessary work, or be fired or required to move without notice from one desolate house in poor repair to another. Typically, such cowboys work when the boss decrees, including evenings and weekends, leaving little time for their families.

When Darcy Lipp-Acord's work outside the home and ranch became necessary to help support the family, she swapped jeans and boots for suit jackets and heels and immersed herself in the education profession. When she could, she took time from her career to focus on raising her children, choosing to relearn her housewifely skills. All the time she's been "just a stay-at-home-mom" or "just a working mother," she's been paying attention to the world as well as to her family's concerns, taking notes for this book. Becoming a mother increased her appreciation of her grandmothers, but she also tested herself against their fears and her own. And always, she challenged herself to move forward in a life filled with beauty and tragedy.

> *I long for the words I write to reach far-away souls, for the stories I tell not to fade away. But with each day, each year, each child, taking time to write those stories becomes harder, seems more selfish.*
> *— Darcy Lipp-Acord*

Lipp-Acord refuses to buy a bread machine because she enjoys the peaceful contemplation of kneading dough as she considers the effects of fertilizers on wheat crops and farm ground as well as on her family's health. She writes about farming organically and the contrasting monoculture of industrial agriculture. She writes fearlessly about the often rocky relationships between white farmers and the Lakotas in the area where she grew up.

> *My ancestors no more romanticized the Indians than the Indians glorified the whites. Living side by side erases many of the prettier myths.*
> *— Darcy Lipp-Acord*

She writes knowledgeably about her German-Russian ancestors who passed down their survival skills along with the knowledge of what it meant to be marginalized, like the Indians, and to lose their land without compensation. She is aware how tourists and city folks who have never made a living in agriculture love to look at open prairie filled with bison and predators such as wolves and mountain lions and imagine an idyllic landscape dotted with tipis. She understands the tragedy looming as ranchers and farmers who have earned a precarious prairie living for three generations are forced out of their homes through the harsh truths of economic reality, their lands to be occupied by buffalo instead.

She writes about disillusionment with her husband's profession and their attempts to find work away from ranching. Carefully, she considers the benefits they enjoyed while Shawn worked for a telephone company and in methane-well reclamation. But she balances the better hours, higher wages, and additional family time against the loss of their horse-raising business and a lifestyle that will allow them to have a big garden, open space for children, and dogs and cats. Living in the country creates more work for her, but she believes it's better for her family. Always, she looks for and finds beauty and joy in her surroundings.

> *There are many ways of owning land. . . . No piece of*
> *property belongs exclusively to me, or to my husband. But*
> *we own the places where we've lived in other ways: we*
> *own the small square of ground in the Boulder River valley*
> *where we buried our dog . . . the old homestead site where*
> *we made love under the cottonwoods.*
> *— Darcy Lipp-Acord*

Although Lipp-Acord writes in her own unique voice of life in today's farming and ranching West, she poignantly represents many young women who are raising families on the land. Whether they own their place, are buying it, or are temporary employees, they form yet another generation of women who don't have the time to write their own stories. Lipp-Acord is one woman, but she tells a dozen stories, her ancestors' voices mingling with her own: the farmer's daughter, the Catholic woman, the wife, the mother, the artist. She is independent, and yet her beauty and grace arise as much from choosing to support her husband's choice of profession as from her own talent. She laughs, cries, prays. She endures.

> *And most of all, I pray that I will have the grace to dance*
> *backward, in order to move forward.*
> *— Darcy Lipp-Acord*

She may feel guilty telling her children to play while she writes, but she keeps writing—between part-time or full-time jobs, between making meals and doing laundry and suffering bouts of sadness. She knows her house is not as sparkling clean as her grandmothers might have wished and feels guilty sitting in front of her computer. She worries about doing her duty to her family, to her church, to her husband, and to herself. She is not always optimistic.

> *I see in my future nothing but the never-ending routine*
> *of home and childcare that I watched the women in my*
> *family live through. Not words, not art, not writing.*
> *— Darcy Lipp-Acord*

These thirteen essays were crafted during more than ten years of time stolen from things the writer knew she "should" do. She grabbed any time she could to write a few lines in longhand to revise later on a laptop computer. On one ranch, she wrote on a rickety table next to the washing machine in the mud room; on another, she used a desk in the bedroom. In coffee shops, she nursed a single cup to scribble for an hour; she wrote sitting in the yard swing while the kids played or at a table in a hamburger joint while they were at preschool. With the children down for a nap or busy with toys, she wrote. In eleven years, she has been pregnant seven times and delivered six children. Through disagreements and diapers, through motherhood and moving, she kept writing. When she gave in to depression, her writing desk became "dusty, untouched for weeks." She yelled at her oldest daughter and drove off to a bar to think about an argument she had had with her husband. She doesn't pretend the work has been easy or that she has done all her chosen tasks without fault, but she keeps writing.

She describes her son's premature birth, her being in his hospital room, miles and hours from her husband and their other children as the television screen shows an airplane hitting the second tower of the World Trade Center. The hospital announces a disaster alert, but she stays beside Cody's bed, resolved that "if the unthinkable happens, our son will not die alone." She lists and analyzes her disagreements with her mother-in-law, discovering comfort in their similarities. She writes about her fear of horses after a childhood accident, decides she was unfair to the horse, and forces herself to learn to ride again, discovering a kinship she had almost lost. Tough and tender, she considers the truths that only women and horses know. Despite disruption, death, depression, she finds joy in her life and her work and she keeps writing. The essays blend together as smoothly as if they had been written in one sitting.

> *Women are more often forced by nature to let go of the*
> *old to make way for the new. We let go of our girlhood to*
> *become women, let go of our bodies to become mothers, let*
> *go of our youthful beauty to become elegant, graceful, wise*
> *old ladies. We let go because we are strong enough to do it.*
> *Because we have to.*
> — *Darcy Lipp-Acord*

As she matures, Lipp-Acord finds herself "circling back" to the way of life that first grounded her, to the lifestyle and religion and places of her elders. She considers how children may accept their parents' ways out of respect and obedience; growing up, they try other ways, walk other paths, and, inevitably, often begin to understand and to honor the choices their ancestors made. At last, she admits that she belongs in the country in the same way her grandmothers did. Yet her choice of country life differs from their choices; her daughters will learn lessons like how to make jelly, but they will also learn by watching her craft stories instead of sweaters.

> *Perhaps they will know that women can do more*
> *than childcare, housework, and handiwork. Perhaps*
> *that will be my impact, my purpose in this world.*

Perhaps that will be enough.
— Darcy Lipp-Acord

Darcy Lipp-Acord's life, and those of the ranching women she represents, differs radically from those of her ancestors, but because she is as strong as her grandmothers were, molded in their image, she will persist. She writes the everyday stories that testify to how she and other western women are living in a world that remains largely ignorant of them. She shows us how their cowboy and rancher husbands and their families survive real physical dangers as well as the personal and economic disasters that are at least as difficult to bear in ranch country as in any city. She holds her family together with one hand and scribbles notes with the other, recording how these families love and laugh even when life is harder than they could have imagined. Women like Lipp-Acord draw strength from their foremothers, their churches, their communities, and their families. Most of all, they look to the land for succor.

> *That voice congratulates me on being a "good" woman because my house stays tidy, my children stay clean, the garden stays weeded. But like the west winds that suck all the moisture from the grass, this voice dries up my joy, my desire, my life.*
> *—Darcy Lipp-Acord*

As this book begins, Lipp-Acord stands in a cemetery filled with family, wondering where she may be buried. By the end, she may not have chosen a resting place, but she knows how to find salvation in the realities of a life that may look impossibly difficult to others. She hasn't climbed the mountains, discovered a cure for disease, or held high office. Instead of choosing to follow her own inclinations, she's chosen to be a responsible wife, a conscientious mother and housekeeper—but she still carries a full complement of the familiar Great Plains guilt. Rural prairie people must drink that guilt in our mothers' milk and eat it in our free-range eggs, or hear it in our church rituals, or acquire it along with our work ethic. Guilt cripples some of us. Darcy Lipp-Acord has transcended it, turned it to beauty. *Circling Back Home: A Plainswoman's Journey* reflects the life of a ranch woman in all its prismatic variety. In a life that some might call ordinary, she has found, and gives to her readers, deep wisdom.

LINDA M. HASSELSTROM
Windbreak Retreat and Hasselstrom Ranch, 2012

PREFACE

From the hillside near my dad's grave, looking northeast, I can see the top of the circle of broad-leaved cottonwood trees in the pasture of our family farm, see the broad, long back of the butte just south of it. Were the cemetery just a mile closer, I could also see the green roof of the old farmhouse, with perhaps a hint of graying red barn peeking through the windbreak of elm and Russian olive. The old farmhouse still stands, but my mom lives in the newer modular home she and Dad purchased in 1980. To my southwest, one row back from Dad's grave, are the graves of my paternal grandmother and grandfather. My mom's parents lie a little farther in that direction, down the gentle slope.

I turn fully east, my eyes tracing the line of gravel county road that runs another eight miles to Glencross, South Dakota, the tiny German community near our farm, where my dad's family settled when he was fourteen. Glencross lies three miles off the main highway, accessible only by three gravel farm roads. It was once a prosperous community, but I've known it only as home to a post office and a liquor store where the local teenagers used to try their luck each Saturday night. Behind me sleeps Timber Lake, the town of my birth, my schooling, my catechism. Boarded-up windows dapple the three-block-long business district; the dozen or so remaining businesses cling to a borderline existence.

This is a good place to be buried, here on a hill overlooking the three places that have formed the center of my dad's, my grandparents' lives. This is their territory, their chosen place. But it is not my territory; it is not where I will be buried.

During my adolescence, I resented Timber Lake its dry, dusty streets, its plain, small-town life. Why did this have to be my hometown? Why did my roots grow from here? In my community, I could see nothing but a few generations' worth of families clinging in quiet desperation to farms too small to make money, to a way of life that balanced them just on the brink of poverty, to traditions so old they were archaic. On the prairies around me, I could see nothing but flatness, dry grasses, buttes no higher than a moun-

tain foothill. The lives of my dad and grandfathers slowly cranked through the yearly routine of spring plantings, fall harvests, and winter blizzards; for my mother and grandmothers, the even more tedious monotony of daily living, from laundry on Monday to church on Sunday. In my grandmothers' territory, life centered on such seemingly mundane questions as "How are you feeling to-day?" "Did he get out in the field?" "How does your garden look?" "Is the baby sleeping through the night?" and the ever-present "Do you think we'll get any rain?" Rarely did I hear discussion of im-portant matters—politics, art, world events. Embarrassed by both my background and my geography, I moved farther west, where craggy peaks and fertile riverlands replaced brittle prairies and dried-up reservoirs.

I live now on the edge of the prairie of northeastern Wyoming, in the shadow of Cloud Peak in the Big Horn Mountains—approxi-mately three hundred miles directly west of Timber Lake. Before that I lived at the base of Togwotee Pass in Wyoming's Wind River range, near a mountain village at seven thousand feet. Before that, I lived on the Missouri River bottom in the middle of the high plains of Montana. And before that, I was six years in and around the Bull, Tobacco Root, and Big Belt mountains of the Montana Rockies.

Though I cherish my memories of these places, none of them has grasped my soul with enough strength to make me say, "This is where I will die. This is where I will be buried." I envy my ancestors that. Both my father's and my mother's parents chose to end their newlywed wanderings with homes near Timber Lake; my dad and mom, other than a tour in Korea for him and a brief time in college for her, have never moved from there. It was, and is, their home, the home of my family, in the most profound sense of the word. But, still, it is not my home.

Early one October, my husband Shawn and I watched a black bear wander through the Wyoming ranch we manage, passing near our house and the corrals. It rested for a while under some cottonwoods, then continued on through a band of sheep and a small bunch of weaned calves. The incident stopped all work at the ranch, and on the neighboring ranches as well. Bears don't typically travel through this valley.

My family and I watched the bear from the moment we learned of its presence until we finally lost sight of it in a draw near another neighbor's home. Not only were we concerned about the ranch livestock, but we were also curious: What was a bear doing here? Though we can see the Big Horn Mountains from our house, they lie over fifty miles away. Bear habitat these days is up there, not out here where the prairie begins. We speculated that this one was an older cub that had lost its mama and was now journeying to find her.

But we were wrong. Shawn accompanied state game wardens who later arrived to relocate the bear. When they found it, they discovered that it was really an old female. She had lost all of her bottom teeth, and her eyes were sunken from dehydration. In fact, the first tranquilizer dart from the warden's gun had no effect because, instead of hitting the muscle in her shoulder, it hit only bone. After examining her, the wardens decided not to try to relocate her—she would die within just a day or two—and so they euthanized her in the draw behind the stackyard, a quarter of a mile from my house. "She came to the prairie to die," one of the wardens said. And she did.

Why would a dying female bear make a fifty-mile journey to a place far away from her natural habitat? Did she make the trek instinctively, or did she simply begin following the creeks, looking for comfort from her pain, and end up here? Could it be that deep inside she knew her ancestors had once roamed both mountains and plains? Could she have wanted to see the plains before she died? I envy the bear her determination, her surety. How did she know that she wanted to die on the prairie?

I do not know where I want to die, or in what land I want my body to be buried. Due in part to my wandering nature, and in part to my husband's career, I am not rooted to any one place. Certainly I will not rest in the same prairie cemetery that overlooks the butte south of my family's farm. I will not rest beside my ancestors.

My husband tells me the exact location of his grave isn't so important to him, as long as he is in the mountains. He points to the range thirty miles outside our windows and says, "That's my home," and I know he doesn't simply mean the Big Horns, but also the Tobacco Roots, the Big Belts, all the ranges of the Rocky Mountains. The mountains are his grounding point; his family's roots lie deep in the high-country ranches of northern Colorado and the small mining towns of western Montana. That he now lives on the edge of the prairie with me hasn't changed who he is essentially: he is mountain bred and raised, a man who is more comfortable tracking elk and leading a pack string than navigating city streets or making business contacts.

In that way, we are alike. Although I attended college in a medium-sized city and even made plans to work in an urban center, the soul that wakes up in my skin every day feels most at ease hiking with my dogs or digging in the soil. My professional life immerses me in public service, education, suit jackets, and heels; therefore, my personal life necessarily includes grass, sky, blue jeans, and bare feet. I know now, in a way I didn't as a teenager, that I belong in the country, the same way my grandmothers did. Even when I return to Timber Lake, I would rather walk the windbreak Dad planted with his brother or sit on that hillside beside his grave—as I've been doing for more than ten years now—than attend my old church or go to the bar with old friends.

At the time I left South Dakota, when I enrolled in college, Dad was facing the end of his life. The emphysema the doctors had diagnosed in him when I was twelve slowly worsened, so that by the time I was eighteen, Dad used oxygen tanks about sixteen hours per day to help him breathe. I had faced death throughout my adolescence. At the brink of adulthood, I ran from it, ran from my background, to a world where beauty so captivated my attention that thoughts of death were easily ignored.

Now, having lost my dad and all four of my grandparents, having birthed my own children, the territory of my grandmothers no longer seems so mundane. The daily, monthly, yearly cycles of birth, death, rebirth: these form the soil of my childhood, provide the wisdom with which I am learning to face the seasons of my own life. The questions that then seemed so trivial—questions about health, crops, weather—I now recognize as questions whose roots delve deepest into the well-being of ourselves, our loved ones. They are the questions that lead us home.

Yet where will home be for me? I long to live in the territory of my grandmothers. Not literally in the physical place that is north-central South Dakota, but in that spiritual place where the cycles of land and life hum with familiarity, where the lilacs planted by

the grandmother are harvested by the granddaughter, where one's lover lies beside one each night for thirty or forty years and then lies on the hillside east of town after that.

Not knowing where I will finally lie unsettles me. I lack a grounding point, a place to die. I embrace and flirt with the mountains nearby like an unfaithful lover, ashamed of my disloyalty to my homeland. But I do not feel the security Shawn feels in the mountains. They frighten me sometimes; they still seem too foreboding and mysterious to be home to me. And yet, I feel too much passion for them ever to return to the plains. The prairie is not the landscape my soul has been searching for all these years.

Or is it? More than a decade after I began my westering adventure, I find myself living in land somewhere between the awesome mountains and the peaceful plains. To the west lie the peaks that are my husband's home, the place of his family, my adopted place for years now. I cannot imagine my sense of beauty, or adventure, or awe being satisfied without them. But some mornings, I gaze over east, finding comfort in the familiar prairie grasses and the hint of humidity that blankets my skin. These are sensory pieces of my past, pieces that my soul recognizes.

And in this surprising, somewhat unwelcome comfort, a question comes: Could my roots lie in precisely the right place for me? After all my rambling, is a return to the edge of the high plains a chance for me finally to send out new shoots in territory that is not so different from my native soil? A chance to live the cycles of a life that I once rejected? And if I do indeed claim this new-but-old landscape as home, as the place of my soul, am I not then obligated to make peace with Timber Lake, with the prairies of South Dakota, with the stories of my ancestors who chose to live there?

As I enter my fourth decade, I am circling back, like a flower that spends its life cycle going back to the seed from which it came. I circle back on the roots that first grounded me: the lifestyle of my ancestors, their religion, their chosen place. I no longer accept their ways simply out of respect and obedience to my elders; I have had to try other ways, explore other choices. Only now, as I begin to be at home in this place not so different from my childhood home, the territory of my grandmothers, do I begin to honor my ancestors' choices. Only now can I honor their stories.

My own stories are part of that honoring. The essays presented here encompass a seven-year time period, from the time my husband Shawn and I moved to the Buffalo, Wyoming, area, where he managed two different ranches, to the time we moved to our present home on a ranch near the Montana-Wyoming border. Because the essays are thematic, they do not follow a strict chronological order. In the beginning essays, however, readers will meet our three oldest daughters: Laura, Carmen, and Maria. As they proceed through the book, our family will grow by three more children: Cody, Emily, and Katie.

In addition to our own family of eight, the book is populated by the characters of my own childhood—my grandmothers and grandfathers, aunts and uncles, parents and siblings—the people who shaped me and my ideas about the importance of our fragile connection to the earth and to one another. Memory is likewise a fragile thing. These are my remembrances, written as truly and clearly as I can render them. Another writer, even one from my own family, might depict them differently. I have made every effort both to relay the truth as I know it and to honor the characters within these essays, for they are the people who share my life. This book is written for them.

You didn't like me at first.

Actually, maybe you did, right at the beginning, before I became a permanent part of your son's life. The evening you and I first met, when I spent an hour playing dolls with your granddaughter, you told Shawn, "You know, not just any girl would do that." But as our relationship deepened, you seemed to change your mind about me. You were irritated when Shawn and I traveled to South Dakota at the end of that first summer so that he could meet my family; you were angry when we announced our engagement that fall. And when I showed you my diamond solitaire engagement ring, all you said was, "Be careful you don't get that caught on something, sticks up so far."

My attempts to earn your approval during those first years of marriage fell short. When I broiled steaks during one of your visits—steaks you brought with you, along with boxes of groceries you figured I wouldn't have—you came running from the living room to check them, sure that they were burning even though I was standing next to the stove. Your remarks toward me often teetered just on the line between a tease and an insult, making it difficult for me to determine your meaning. While most of your barbs targeted my cooking skills, still shaky in those days, you also critiqued my career choice, my religion, my parenting techniques. When our first daughter, Laura, was born, you approved of my nursing her . . . until she was six months old, at which point you and Butch, Shawn's dad, started "joking" that she would enter kindergarten unweaned.

More than not understanding my ways, you seemed to resent my place in Shawn's life. When you visited us in our Boulder, Montana, trailer house two months after the wedding, you cried when Shawn snuggled up with me on the loveseat instead of sitting next to you on the couch. The week when your mother was dying in Helena, you and Butch stayed with us in Boulder. One day you lost your temper with Shawn because, instead of riding up to Helena with you, he decided to wait for me to get off work before driving up to see Grandma Patrick. And the following autumn, when you, Butch, and my younger sister Michele joined us for Thanksgiving dinner, you were angry that Michele rode with us on a trip back to visit Grandma's grave in Deer Lodge.

As I look back now, though, I see that perhaps there was more to your resentment than simply not wanting to let go of your youngest son, the one you always referred to as "the fun one." There was more to it than having to make room for me and my family in your life, as well. You knew that only a certain type of woman would be happy married to Shawn, and I don't think you thought of me as that type. Once, he told me that you wanted him to continue to date a certain ranch girl, even after he'd met me; perhaps you thought her background matched Shawn's better than mine.

For Shawn is a cowboy, pure and simple. When you and I met, I was a college student working for the summer at a guest ranch. My majors were English and Spanish, my passions writing and foreign cultures, my career goals journalism and international relations. Not exactly the cowgirl you envisioned for your son, was I?

I am a ranch girl, though. I grew up on a small farm/cattle ranch in South Dakota, and the rhythms of agricultural life ring familiar to me. The critical difference between my youth and the early years of my marriage is that my dad owned our operation. Shawn and Butch have always worked as hired hands on someone else's ranches.

My own mom taught me the skills I'd need to be a rancher's wife. Whether you believed it at first or not, I've always been capable of managing a ranch house and family. Mom taught me how to cook meals, how to tend the garden, how to run for parts, how to watch the sky. She instilled in me a love for the country and for country living, and a firm belief that no better lifestyle exists. She showed me a thousand times over the importance of woman's work, on a ranch or anywhere, and she gave me the confidence to know that I could be a good wife and mother. Just as importantly, after my dad was diagnosed with emphysema, she urged me to be prepared to support my children and myself if a husband could not, and she is the reason I have my degree today.

However, prepared as I thought I was, the first five years of my marriage to Shawn yanked the rug of self-confidence out from under me. Mom wasn't able to teach me that being the ranch *hand's* wife placed me lowest on the scale of importance on most ranches. My skills in and out of the home meant little in the face of suppers turning cold on the table, newborn calves dirtying the bathtub, Sunday afternoon plans disintegrating in a single phone call from the boss. I think you sensed how much difficulty I would have in dealing with these realities.

Even worse than the inconveniences were the feelings of unimportance and helplessness. From those first years, I remember waiting by the living-room window, peering at the headlights of every passing vehicle on the highway, hoping that Shawn's would turn into the driveway soon. So often, he went out on jobs alone. I learned to memorize where he would be riding each day so I could go out looking for him if he wasn't home by ten that night. As half past nine and a quarter to ten came and went, I remember praying desperately that he was still alive, knowing that if I lost him, I would lose not only my best friend but also the home that went with the ranch job.

These frustrations so overwhelmed me on the Boulder place that when a job offer from your hometown came through, I agreed to move, even though I was reluctant to live closer to you and Butch. True to the nature of ranch help, though, we believed that the next place would be better, that the next boss would be easier to work for. Was that the way you and Butch thought as you moved through all those jobs, as well? Because it proved to be a faulty thought pattern: if anything, the second job was harder, and I left a full-time teaching job for one that was less than half-time, so we were poorer than before, too.

A year into that second job, our second daughter was born. That was the summer when things went from bad to worse, the summer when Shawn nearly drowned crossing the Missouri River on horseback by himself, the summer when he had to cut nearly all the hay crop with the dew, working from nine at night until he couldn't cut anymore. Carmen happened to be born during a rainy week; after the rain stopped, Shawn was so busy with the ranch that he had little chance to see her awake until she was six weeks old. I began looking for ways out of this lonely, bleak lifestyle.

Curiously enough, you and I began forming a fragile alliance that summer. You began to help me with the girls without criticizing, and I began to see that your brusque manner concealed wisdom earned from years of living a life like mine. Still, Shawn and I decided our marriage would not sustain another summer like that one, and in a move that both surprised and hurt you, Shawn decided to quit ranching. A teaching job drew us away from you to Wyoming; I became the full-time breadwinner in the family while Shawn decided on a new career.

Shortly after our move, the local telephone company hired him to do construction work. Suddenly, he was working a predictable schedule, with weekends off, great pay, and good benefits. Coming from a ranching background, this type of job seemed to be part of a dream. I could plan weekends and vacations; we had supper together every night; and our bank account looked relatively healthy, even though our family now included three young children.

But you probably knew Shawn wouldn't remain happy in that job. He was satisfied for the first year or so, but telephone companies just don't use many horses these days. He again took to spending more time away from home, either training colts or shoeing horses for neighbors—anything to capture the smell, the feel, the companionship of horses. Now, after two and a half years, a wealthy landowner has asked Shawn to manage a new ranch he is purchasing near the Big Horn Mountains in northeastern Wyoming. Neither of us is considering turning down the offer. Shawn is simply so unhappy, Ann; not even my anxieties about returning to ranch life would convince me to ask him to stay.

And so, nine years into a relationship that has been tenuous at best, I am turning to you for advice. You have played the role of the ranch-hand's wife with grace and humor for as long as I've known you, and probably for most of the forty-odd years that you've followed Butch from ranch to ranch. How do you do it?

Do you ever get angry? On our first wedding anniversary, Shawn had to work late to fix a broken-down combine. The romantic dinner I'd prepared turned cold and wilted. I threw a bona fide temper tantrum, screaming, crying, even throwing a bowl against the wall. I couldn't believe the ranch was more important than our wedding anniversary. Years after the episode, Shawn told me he remembered you getting all dressed up on your anniversary, then crying when you realized that Butch was working late. Of course you cried, but do you ever rage? Does anger ever completely overwhelm you? Do you ever curse the ranches that take your husband away from you so much of the time?

Do you ever curse the bosses? One boss we had made a regular appearance in church every Sunday. I know, because I watched him from the cry room, where I was struggling with two toddlers by myself. When we interviewed with him, he told us that family and religion were "the most important things." He must have meant his own family and faith, because guess who was doing the chores every Sunday morning?

Do you ever grow accustomed to not knowing when your husband will come home? Do you still panic when Butch is out much later than you'd expected? When Shawn and I were first dating, you told me you'd learned not to expect him "until I see the whites of his eyes." How did you cultivate such calm?

And how do you calmly accept that your home might cease to be home in the most unexpected ways: if your husband is injured or killed, if the boss becomes unhappy, if financial mismanagement—not your own—causes the ranch to go under? How do you continue to move from ranch to ranch, not knowing where you will end your life? Your children can no longer return to the home in which they grew up, because you no longer live there; doesn't this trouble you? I know that, in facing your diabetes, you have had to think about your final resting place; you have decided you want to be buried in Deer Lodge, next to your family. Butch has decided to rest there, as well. I find it ironic, and morbidly funny, that, after all these years of following him, you will make him follow you to the grave. But I won't rest with my parents in Timber Lake, nor will Shawn. My most important question is one that I think you have not yet managed to answer for yourself: How do we walk the balance of this ranching lifestyle in a way that allows us to put down permanent roots, to make a home at last?

I hope you can provide some answers, Ann. Most of all, I hope you will not tell me that there are no answers. My biggest fear is of giving up, of having to accept the difficulties of this lifestyle because I have no power to change them. But I don't believe I am powerless; neither are you. Something—love for a man, love for a place—has drawn us both to this life. There are lessons to learn here, and I hope you'll share what you've gleaned from your journey.

From my childhood, and from those first five years of marriage, I know this way of life holds more potential for joy and for heartache than most others. Teach me to find the joys, to see the rewards behind the hardships. I expect the best reward will be the look of bliss on Shawn's face as he saddles his horse in preparation for the day's work.

But you already knew that, didn't you?

Boy, it's awfully wet to be planting a garden today," observes Bert, one of our new neighbors, when he drives in to see my husband. My family and I are spending this misty May morning planting corn and potatoes in the freshly tilled plot next to the barn. Until Bert's truck rounded the corner of our driveway, we heard no noise here next to the ranch buildings, away from the main road.

I stand, look down at my dirty jeans, and shrug. "It's nearly the end of May. This corn should have been in weeks ago. And the potatoes we should have had in the ground in April."

"April! It was snowing in April," he laughs.

"Well, the old Germans where I grew up used to say to plant potatoes on Good Friday, and that's in April."

"Farmer's daughter," Shawn grins, and they share a chuckle.

I don't care; a little rain never hurt me. "Never cuss the rain," Dad always said. And today, it's only spitting a little.

"Momma, show me where to put more seeds. I wanna plant more corn," begs Carmen, my three-year-old middle daughter.

"OK. Laura, do you still want to help?"

"Uh-huh," our oldest daughter nods.

So, I drop to the ground and resume poking my finger into the wet soil. After I make the holes, Laura and Carmen drop the dried corn kernels in and pat the dirt back into place. Maria, the youngest, is eager to help with the patting. "Go soft, Maria, not so hard," I warn. Although it's not exactly mud, the dirt is moist enough to cake the palms of their little hands. *Good,* I tell myself, *the seeds will germinate better that way.*

I have a moment to think as I'm poking holes. *Farmer's daughter?* Not long ago, that comment would have set me on the defensive. When I left the farm in South Dakota, I had no intention of going back to either the place or the way of life. My parents seemed to accept that I had chosen a path different from theirs. Dad drove me to college in Montana, leaving me in a place farther west than any of his children had gone. Both he and Mom encouraged my studies, expressing nervousness about my travel plans only when the destinations seemed risky: El Salvador, Mexico, Chile. Dad's eyes sparkled as I spoke of far-off places. His only traveling had taken him to Korea to fight in the war, but he read every issue of *National Geographic* that came to our house, and nobody could beat him on history or geography questions when we played Trivial Pursuit. Mom occasionally wondered aloud how often I'd visit if I was living so far from home, but guilt didn't even deter my plans. The summer of my younger sister's wedding, my siblings and I joked that I'd be the only old maid of the five of us.

Then, I spent a summer in West Yellowstone, Montana, and entered the autumn in love. From the instant I knew I loved Shawn, I also knew he'd never fit into the life I had planned. Cowboy to the core, he was more like my dad than I wanted to admit. When I brought Shawn home, after a lineup of boyfriends with earrings, fast cars, and maturity problems, Dad seemed more relieved than surprised. "Wish I had some fence that needed fixing," was all he said. Believing that I belonged with Shawn more than in any foreign embassy, I took classes to earn my teaching certificate and spent years passing my love of literature and cultures to teenagers as Shawn's work took us from ranch to ranch, until we finally left Montana and moved to Wyoming. And I bore three children. I look at them now, carefully patting the black dirt over golden yellow seeds. It seems a good trade.

"How's it goin'?" Shawn returns to the garden as Bert honks, waves, and drives away. Shawn's black cowboy hat drips rain off the brim; my hair is matted to my head.

"Almost done."

He crouches and tosses a clump of grass over the garden edge. "I really should have tilled this twice, but I didn't have time. All this rain."

"Hey, for the first garden we've had in six years, I'd say it looks pretty good," I remind him. This ranch in northeastern Wyoming is our seventh home in eight years. After a three-year hiatus from ranching, Shawn left his "good" telephone-company job last winter to return to the cowboy life: this time, managing this place for the owner, who lives in a city two hours away. "Besides, next spring this plot will look even better," I add hopefully.

"You bet," he brightens, looking around at the little red house, the barn, the corrals. "I think we'll be here a while. Well, guys, looks like you're about done. What say we wash up and get some lunch?"

"Yeah!" the girls shout, and run for the house. I linger a few minutes, watching a group of wild turkeys head for shelter in the willows near the creek. As I watch, my memory wanders back to planting season on the farm in South Dakota.

Dad raised corn, sunflowers, wheat, and other grains there, along with a small herd of cattle, and, until I was about six, pigs. I remember wandering into the kitchen on sunny spring mornings to find it nearly empty, only Mom there working, my older brother and sister already at school. "Where's Daddy?" my little sisters and I would ask sleepily, much the way my girls ask for their dad when he leaves early these days.

"Already out in the field," Mom would answer.

Dad did the planting alone, for we could afford no hired help. The only crop we sowed as a family was potatoes—just a few furrows for our own consumption. We always chose a drizzly, wet April afternoon when there was a break in the other spring work. Dad used the tractor and plow to break the soil, and Mom, my brother, sisters, and I followed along on foot, dropping seed potatoes into the rich, moist, sweet-smelling dirt.

My childhood mornings in the spring revolved around playing with my little sisters in our room upstairs in the old farmhouse, the same room where Dad and his younger brother had slept as boys. When we tired of that, we played outside on the swings, or at Mom's feet on the kitchen floor.

As we played, our ears perked every time the drone of the tractor got louder. "Daddy's coming in!" I'd tell my sisters, and we would run to the window or door to watch for him. But then he'd turn the big orange tractor and start down the next row, the drone fading away once again. After a while, we quit listening.

At lunchtime, Dad would come into the house, smelling of tractor grease and sweat. He always ate quickly. "How's everything going?" Mom would ask.

"OK, I guess," he'd say, or, "I had to run to town for a part; that took nearly an hour." They, too, always spoke about farming. Always the crops, the weather, how far along the neighbors were with their planting. Often, they ate together silently, listening to the noontime market reports on the kitchen radio. Then Dad would light a cigarette, put his straw hat back on, and be out the door before I'd eaten half my lunch.

The rain worsens, bringing me back to this place, this time. *So far, so good*, I think. At least Shawn helps put the girls to bed most nights, and he sees them often during the day. We've been here three months, and he's been able to keep his working hours reasonable. And on the weekends, he can take time off to do family projects like planting this garden. *Maybe this place will be different from the others. Maybe we can make ranching work for us this time.* I look at the black dirt drinking in the drops. Satisfaction fills me: the corn is in the ground.

It's now two months later, a perfect July morning, so I decide to work in the garden; the carrots really need weeding. I walk across the driveway, noticing that the cornstalks stretch almost to my head now, reaching long stiff leaves out to absorb sun and to shade the ground below. I planted peas and pumpkins in the same space, following modern companion-gardening advice. Actually, the practice is not so modern: I've read that Buffalo Bird Woman, an elder of the Hidatsa tribe of North Dakota, remembered planting the same way. The beans, or peas, enrich the soil at the base of the stalks with nitrogen, while squash or pumpkins thrive in the shade of the taller plants' leaves. Makes sense. Still, I'm sure Dad would chuckle at my gardening design. Peas and pumpkins never entered his cornfields.

As I near the corn, I spot soft, feathery tassels rising from the tops of the tallest center stalks. Tassels! This means the corn will pollinate and bear fruit! Abandoning all plans to weed the carrots, I drop to my knees amid the stalks, ruthlessly pulling up stray blades of grass. The fragile, thready stems of pea plants climb up some stalks, but I see no sign of the pumpkins yet.

My daughters, who have tired of playing inside, call out from the front steps, "Mom! Can we come out with you?"

"Sure, just put your shoes on. And help Maria."

One at a time, they slam open the gate and scramble out into the barnyard. Laura goes to the machine shed to find her bike; Carmen pushes through the stalks to talk to me; and Maria chases our new puppy, Sam. A year ago, they had no safe, open space like this to play in. Our yard then consisted of clumps of sagebrush, gravel, and cacti, so they spent much of their summers inside. Here, I can't keep them in. They can play in our fenced backyard, ride bikes in the driveway, or visit the horses. And because I'm with them all day, instead of entrusting them to a babysitter, I don't worry as much about the dangers of the creek or the livestock. As I pluck out grass stems, I reassure myself that our decision to return to ranch life was the right one for the girls, even if Shawn's hours are getting longer now that summer's here.

I know that we kids were a big reason Mom and Dad continued to operate the family farm, even when the lifestyle drained all their energy and finances. Never once during my childhood did I go to day care; Dad hired a neighbor girl to stay with us whenever Mom went into the hospital for her various health problems. The only other babysitters I knew were my grandmother and an aunt who lived nearby. Never once during my childhood did I have to stay inside to be safe; our yard was perfectly safe, removed from the highway by a mile of gravel road and protected by several family dogs. Only occasional rattlesnakes and rabid skunks forced us to stay away from the tall grass and one of the more decrepit farm buildings. Never once during my childhood did I move, have to start over at another school, another town. The farm kept us in one place, a place where we knew our every neighbor, schoolmate, and relative practically from birth. The farm was my security blanket. For many years after I grew up, I scoffed at that security, thrived on the change that moving brought. Now, as I consider my children's needs over my own, I see what my dad and mom knew: that security is necessary for a child. And that kind of security is exactly what we came here to give our girls.

Mid-July heat shimmers up from the ground, even though it is barely seven o'clock in the morning. *I just weeded this corn*, I think as I dutifully turn the sprinklers on. But, already, green blades of grass poke up, refusing to give up their claim on this recently tilled patch of ground. Persistent little plants. I sigh and walk back to the house. The air is dry and warm; there wasn't even a dew last night. These last three weeks have been intensely hot, with no rain. The nonirrigated ground is parched; I feel sorry for dryland farmers, like my dad.

Shawn watches for me at the door; he stays in to care for the kids during the early morning so I can go for a walk. I suppose he's ready to go to work. Because of the hot, dry summer we've had, his boss wants him to keep the irrigation sprinklers on constantly; and because the sprinklers are old and in disrepair, Shawn spends nearly all his time babysitting them.

"Hi," I say as I come in.

"Hi. How was your walk?" He doesn't look up as he tugs his boot laces impatiently. From the way his cheek is trembling, I can tell he's grinding his teeth.

"It was OK. What's wrong? Did I take too long?" I ask.

"Oh . . . Bob called."

"What now?" I sigh. Phone calls from the boss, who lives in Casper, Wyoming, are not a pleasant way to start the morning. In the last few weeks, the luster of this ideal job has begun to fade. Actually, there have been problems with the boss since last spring, but I haven't wanted to admit them. I really wanted this arrangement to work out.

"Oh, he's all worried that his alfalfa isn't growing fast enough. Today he said he doesn't think we'll even get a second cutting," Shawn answers disgustedly. "Darc, I told him, way last spring, that we needed to fertilize these fields. You've got to put back into the ground what you take out."

"And the Andersens never did," I say, referring to the previous owners.

"The Andersens never did a goddamn thing, but Bob won't believe me when I tell him that."

"Bob doesn't believe you on anything these days," I say. I understand Shawn's frustration; despite his youth, he has a lifetime of ranching experience. His boss, on the other hand, has none. While there's nothing wrong with trying a new venture like this ranch, Bob won't admit that he knows nothing, and he won't take Shawn's advice. I begin pouring cereal into bowls for the girls, who have wandered out of their rooms while we've been talking. "So, what does he want you to do about the alfalfa?"

"Nothing. He just called to bitch." Shawn stands up. "Well, I better get out there. You know, honey, it's not my fault that it isn't growing."

"I know it's not, and so do you. You've done wonders with this place. Hey, maybe we'll get some good news tonight, huh?"

"I hope so," he says as he kisses the girls and me, then ducks out the door.

Later, as I eat my own breakfast, I stare out the window at the garden. Tonight, we have a job interview for a position with a neighboring ranch. If we take the job, we'll move again, probably in the fall. *God, I don't want to move*, I think. But I can't deny the problems we face. When we left our old jobs to come here, Bob promised Shawn a new house, cattle to run, and complete management of the place. Now, six months later, we still live in a tiny, pre-1960 house; there are a total of five cattle on the whole place; and management of the ranch is subject to Bob's whims, not Shawn's experience.

I look over at my corn, still fighting the grass and weeds and drought so valiantly. Will I even be able to harvest it, or will we move before then? Why did I even hope to have a garden this year? Why was I fooled into thinking this ranch job would be different than any of the others?

When we first married, in 1991, Shawn worked as a hired hand on a Montana ranch run by four brothers. He referred to himself as "low man on the totem pole," and he was right. Shawn and his immediate boss, the third brother, worked six and a half days a week, ten to fourteen hours a day. Although we put up with that schedule for two years, Laura's birth pushed the situation from barely tolerable to absolutely unacceptable. Shawn quit, and we moved to the northern Montana town where he had lived during high school. There, a former employer hired Shawn as cattle foreman. We soon learned, however, that the term "foreman" meant nothing, because Shawn had no power to fire employees who didn't do their work. Instead, he took on their unfinished jobs, and his hours lengthened again.

Finally, after two more years, he did what his family and friends considered unthinkable: he quit ranching. We moved to Wyoming, to a little mountain town where I had a full-time teaching position. For the first time in his life, Shawn had no definite job; never before had he considered not ranching. He did some horseshoeing and leather work and took care of the girls while I taught. After a few months, he accepted a job with the local telephone company. Suddenly, he was working eight to five, Monday through Friday, with occasional overtime, a paid vacation, health insurance, a pension plan, and yearly raises. The irony made him shake his head; he was getting more money and more benefits, yet working fewer hours than he ever had ranching. When Maria was born, a year after we moved there, he was able to take off all the time he needed.

For the first six months, Shawn liked his job, but when the novelty wore off, he was simply bored. He hated doing service calls, spending most of his day in a pickup truck. Laying new telephone cable was a little better, because he challenged himself to bury more feet in a day than any other employee. Still, the absence of animals to care for and, especially, horses to ride, was more than he could stand. He threw his energy into his part-time farrier business, shoeing two or three head a night and taking on more clients just so he could be with horses. We moved four times, searching for, but never quite finding, the perfect country place in which to make our home. Shawn began to be difficult to get out of bed in the morning, preferring to sleep rather than go to work. At night, he'd often be too tired or too frustrated to play with the girls, and he began shooing them downstairs to watch TV while he complained to me about his day.

Then came the offer I think he had pined for. Bob, on whose "ranchette" we had become caretakers, wanted to buy a ranch in northeastern Wyoming—this ranch—and hire Shawn to manage it. I hesitated; I didn't like Bob, and I didn't like ranching hours. But Shawn had been so unhappy for so long. He promised he'd keep the hours reasonable and take weekends off, except for haying and calving seasons. And then came the clincher: Bob would pay him enough so I could quit teaching and finally stay home with the girls.

Now, as my angels are fighting over yet another toy, I wonder how wise I was to agree to this. I'll be alone with them from now until they go to bed at eight tonight. Shawn will be in for lunch and supper, but just to eat, not to take time with the kids. The first cutting of hay was finished two weeks ago, but he's still working twelve to thirteen-hour days. Sometimes I think I've spent my whole life waiting for men to come in from the fields.

On this mid-August evening, Shawn and I stand in the garden. "Look, here's another one!" I exclaim to Shawn from between two rows of corn.

"Yeah, I've found a few more here, too. Looks like we're gonna have a corn crop this year."

I discovered the tiny ears earlier this August evening when I was picking lettuce and carrots for supper. The tomatoes came first: one of the dozens of green fruits finally ripened into a luscious, rich red, and I picked it to add to our salad. Then, not daring to hope for more, I stepped over to the corn. Sure enough, there, on the tallest stalks, tiny ears were sprouting.

Now, after supper, Shawn and I are counting. We find sixty-four ears growing on about half the stalks; others have just begun to tassel.

"Do you think these others will still produce?" I ask him. Some of them reach only as high as my chest and are withered, an effect of little water and the rocky soil at the garden's edge.

"I don't know, Hon, as hot as it's been. All we can do is keep the water on them."

"You're right. Well, I guess I'd better get these kids in the bath; it's getting late."

"You go on ahead; I'll be there to help you in a bit."

As I run water for the girls' bath, I look out the bathroom window to see Shawn carefully adjusting the sprinkler so the spray will hit the far end of the corn. Corn adapts to most soil and climate conditions, making it the most widely grown of all grains. Still, it requires an abundant supply of water. The Incas of what is now South America, upon acquiring new territory, made preparing the land for corn their first priority; they leveled the mountain ground, building an elaborate system of terraces to allow for irrigation. The Aztec tribe of present-day Mexico created a system of artificial islands, or *chinampas*, that used a grid of waterways for both irrigation and transportation. My dad, on his dryland South Dakota farm, relied on Mother Nature. I have the easiest watering: I string a hose from a spigot in the barn to run my sprinklers.

Even with enough watering, though, nature or misfortune sometimes takes a crop. When I was about nine, Dad lost half the corn harvest to a fire. My mom, little sisters, and I had been at my grandparents' house, visiting relatives. As we drove home, we could see a cloud of smoke to the southeast; Mom guessed the fire to be at one of the neighbors' farms. But as we turned from the highway onto the gravel road that led to our place, the sheriff's car and volunteer firetrucks from town followed us. Mom flew down the road, commanding us to pray Hail Marys that the house wasn't on fire. Instead, as we pulled into the yard, we saw a thick black strip running the length of the bright yellow pile of corn Dad had just finished harvesting. He'd piled it on the ground because he had run out of room in the grain bins. Neighbors swarmed through the yard. Mom learned from my older sister that she had burned papers earlier that afternoon, and the fire had escaped the barrel where we burned our trash, run across the yard to the corn, and then spread down to the pasture. As Mom drove us back out of the yard over to Grandma's, I looked back. My dad was standing atop the smoking corn pile by himself, holding an old rag and beating down the flames that had consumed so much of that year's toil.

Now, it's nearly nine o'clock in the evening, and the air is just starting to cool. The kids are in bed, and Shawn and I are sitting on the back steps, drinking beer. The bottom board of the steps is rotted away, and I pick at the remaining wood as we talk.

"You look relieved, finally," I smile, watching him gaze out over the backyard at six whitetail does browsing in the hay meadow.

He grins and takes a long drink. "Why shouldn't I be relieved? He's finally gone." Bob usually arrives on the ranch each week on Wednesday night and returns to his other home on Sunday afternoon. We sit quietly for a few minutes, watching the sinking sun backlight the peaks of the Big Horns pink and gold. When Shawn speaks again, his voice is serious, desperate. "I don't know how much more of this I can take, Darc."

"I know. I don't know how much more our *family* can take."

This morning, as we were dressing the girls for church and packing a picnic lunch for afterwards, Bob called. He ordered Shawn to come over to his house to "go over some things." Shawn asked whether a meeting could wait until later in the afternoon, but Bob cut him short by saying he had to leave the ranch by noon, and Shawn would just have to change his own plans. Even though Bob had been here since Wednesday, he'd hardly bothered to acknowledge Shawn until now.

When Shawn hung up and announced that he'd have to go over to the office, I pounced on him about doing ranch business on Sundays—something he'd promised not to do. He yelled back about having no control over the situation and asked me what we'd do if Bob fired him. I retorted that I'd consider being fired from this place a blessing, and he, of course, took offense to that: we had moved here primarily to please him, not me. So, he slammed out of the house and spun gravel under his pickup tires as he drove over to Bob's. I changed the girls back into shorts—I was way too angry for church—and we went for a long drive in the mountains. Shawn found us, three hours later, at the base of Crazy Woman Canyon.

After we'd both apologized for screaming, Shawn told me that when he'd arrived in Bob's office, he reiterated that meeting on Sundays was not necessary. After all, Bob called him nearly every Monday morning. Bob responded with, "Well, I'm the one who signs the checks, so we'll meet when I say." To that, Shawn answered, "Well, looks like you won't have to worry about signing my check much longer." Then he walked out—no discussion of whatever had been so important—and went looking for the girls and me.

When we got back home, there was an apologetic message from Bob's wife on the answering machine, and they were gone. We spent the rest of the afternoon not discussing the situation; instead, we played with the girls and tried to smooth over the damage done that morning.

Now, beer in hand, Shawn gropes for a solution. Assuming he still has a job with Bob, he realizes he can't stay here much longer, so where should he work? Reluctantly, we begin to realize that we may not be able to perfect ranching. Working on Sundays, working late, and working through the weekends have been problems common to every ranch job we've had. Neither Shawn nor I are strangers to hard work; our own childhoods were filled with it. But questioning our childhoods, and thinking about how we want to raise our own family, has caused us to weigh carefully the difference between jobs that have urgency, and jobs that can wait until Monday morning. We both believe that ranching, like any career, necessitates both types of work.

"Do you think I was wrong to come back to ranching?" Shawn asks me earnestly.

I'm quiet for a long time. I have to try to say the right thing, even if I don't quite believe it. Finally, I say, "No, Shawn, you weren't. It's a good lifestyle, especially for the kids. I just wish there could be some limits on the hours."

"Maybe I should go back to the telephone company."

"But you weren't happy there."

He just shrugs and takes another drink. I continue, "Look, Shawn, you have to do whatever makes you happy. You love ranching, so you have to do that."

"But *you* don't love ranching."

His words hit me hard. There is a truth to them that I don't want to acknowledge; how can I not love a lifestyle I was born into? I've struggled for eight years to be my father's daughter, to be a good wife and friend to Shawn, but I still haven't embraced the way of life they chose. I pick my words carefully when I respond. "I don't love some parts of it, you're right, but other parts I do. I won't ask you to quit, but we have to set some limits. There should be some time set aside that's just for family, and Bob shouldn't be able to mess with it. This isn't even a busy time of year here; why should you have to work on a Sunday?"

"I know—he's just such an asshole. I sure wish those guys would find out if they got the lease on that place." He's been offered the job we interviewed for last month, but the whole position is contingent on the rancher being able to lease another ranch for Shawn to manage.

"I hope so, too, but for now we have to figure out how to deal with Bob." More silence. He doesn't want to make things work with Bob anymore, so I drop that approach. To cheer him up, I ask, "If they get the lease, when will we move?"

"Probably October or November—or it may not be 'til spring."

"Well, at least I get to harvest my corn that way," I respond, only half joking. As I peer into the growing duskiness of the August night, bats swoop after mosquitoes and a chilly breeze picks up. I thought this place would be home for a while, that we could put down roots here. But tonight, those roots—and a permanent home to go with them—are as elusive as the bats darting through the cottonwoods.

Late September sunshine warms the side of the old barn as my daughters and I rustle through the corn patch, searching for unpicked ears. "No, Maria, we're not husking these . . . Leave the husk on, Maria . . . Maria, no!" Our two-year-old daughter toddles purposefully to the foals' pen and begins peeling off layers of husk from the ear of corn she clutches.

"No, honey, we're not doing that today. Put it in the sack." I retrieve the partially husked ear and add it to one of several grocery bags full of corn. Maria, who doesn't understand this sudden change in her routine, throws herself to the ground in a tantrum. Today is the final harvest of my corn crop: Already I've lost about half my tomatoes to a mid-September frost, and the corn stalks are turning brown. These ears won't ripen any more.

Over the last month, my daughters and I have enjoyed the late-summer pleasure of picking fresh sweet corn just before supper, husking it next to the horse pens, and then returning to the house to plunge the ears in boiling water just long enough to turn them a deep golden yellow. The honey-sweet kernels have brightened many weekday suppers and Sunday dinners. I'll freeze some of these last ears to hold onto a little of the taste of summer through the winter; the rest I'll serve at a harvest dinner we're having next week.

As I settle on the front steps of the house to sort ears—Maria now playing happily with Laura and Carmen in the backyard—I reflect on the bittersweet memories Dad must have had during harvest time. In my sweet-corn harvest I've felt some of the thrill and gratitude he felt as he brought in his crops: his springtime faith mingled with his summer sweat and prayers to give back an autumn miracle. And yet, memories of regret and loss must have discolored his satisfaction; after all, it was the corn harvest that took his fingers.

20   I can still hear my mom telling me the story one summer af-

ternoon. She was sewing during my sisters' naptime, and I sat on her double bed, playing with the tin of buttons she and my two grandmothers had collected. She mentioned that Dad would be harvesting soon. I asked her how Dad harvested corn, and she reminded me of the big silver cornpicker he pulled behind the tractor. Although American Indians and early non-Indian farmers picked by hand, by the time my dad's generation was farming, implement manufacturers had successfully developed a mechanical cornpicker. It looked like a gray dinosaur, creeping through the fields, swallowing entire ears of corn and spitting them back out into its bin, stripped of the husks, dark yellow. In its wake stood the naked stalks, left for stubble to hold the soil against the South Dakota wind.

The conversation lulled for a few minutes; Mom must have been debating as to whether I was old enough to know this family story. Finally, she continued, "You know, Daddy lost his fingers because of a cornpicker.

"It happened before you were born, when your brother David was only three. Dad was picking corn in the field by the old pump house, right by Grandpa Quinn's land. You know, where our old mailbox used to be? Well, the cornpicker got plugged, so Dad jumped off the tractor to clean it out. He didn't turn the tractor off first, so the belts were still moving. One got wrapped around Dad's hand, and started pulling his whole arm into the picker."

"What did he do?" I asked breathlessly.

"He had to pull back to try to save himself, and he ended up pulling his fingers off. He probably would have died if he didn't do that."

"Then what happened?" I was shocked, imagining the pain. How could you pull off your own fingers? Up until then, I had childishly believed that grown men lost their fingers naturally, the way I lost baby teeth. Never mind that my uncles and grandpas still had all their fingers; since my dad was missing four, I assumed that his hand was normal.

Mom continued: "He was lucky that the sealers on the cornpicker were so hot they burned the ends of the veins, so he didn't bleed to death. He walked all the way across the field to where his truck was parked. Then he drove himself over to Grandpa's."

"Why didn't he drive home?"

"Because I wouldn't have been here. That was when Uncle Fritz still lived here on the farm, and Dad drove out from town every day. Your uncle was somewhere out in the field, too. Besides, Dad was actually closer to Grandpa's place than here. Anyway, I had been to Mobridge that day, and I stopped to see Grandma on the way home. I got there just after Dad pulled in, and your uncle Dan came running out of the house, saying Dad had cut off his fingers. Grandpa drove us back to Mobridge to the hospital; I thought we would never get there." She said the doctors sewed Dad's hand to his stomach to get skin to grow over the spot where his fingers used to be. If the skin graft hadn't worked, they would have amputated his whole hand.

Fortunately, the graft did work, and Dad lived the rest of his life minus four fingers. The skin from the graft looked different than the rest of his hand: smoother, with no deep furrows. It was lighter colored, too, never the deep brown color of the rest of Dad's skin. Under the graft, I could see bumps where the skin pulled tight over the joints of his index and middle fingers.

I think now of all the tasks Dad completed using just the stump and thumb of his right hand: carrying bags and boxes, using a knife or fork, driving machinery. For tasks requiring more intricate dexterity, he taught himself to be left-handed. Mom told me once that, shortly after the accident, Dad needed a shave badly. Hesitantly—"I was so afraid I would cut him,"—Mom took his razor and began shaving his cheek. After several of her clumsy strokes, Dad snapped, "Here, let me do it," took the razor in his left hand, and began to shave.

The only manual task he struggled with was writing. He generally avoided writing more than figures on farm paperwork; at a store, he'd sign his name in trembling, left-handed cursive and hand the check to the clerk to finish. Once, though, when I was a sophomore in college, he wrote me an entire letter, just from him. I still have it. The short two pages are filled with details of the end of a career: he tells of cutting up old machinery to sell for scrap metal, and of selling off the last of his cows. He talks about going to the veteran's hospital for his regular checkup but says no more of his failing health. He notes that he'd been letting some hunters onto our land, and I explode in teary laughter every time I read his wry assessment of their hunting abilities: "Leo and Tim were here last weekend pheasant hunting. Saw lots of birds but didn't get meny. Cant shoot. Ive had 10 hunters here for the past two Days. Lot of birds. Cant shoot either." Though I only knew Dad to shoot rabid skunks and dying cows, he learned to balance the rifle on his right hand and pull the trigger with his left. And he could shoot.

I watch my own fingers nimbly pull silks from the ears of corn I plan to freeze and start planning next week's party in my mind. In the three years we lived in Dubois, Wyoming, we never had friends over, but this will be the second party we've hosted here in three months. Tears well up as I consider leaving this valley of small, friendly ranches. The stress of working for Bob has forced us into an unhappy decision: Shawn will look for other work while we wait to hear whether the neighbors get that lease. He's already applied for a job with a construction crew that is working in the area through spring. I'll probably do substitute teaching to help stretch the budget. All this means a return to town living: no wildlife, no horse pasture, no room for the girls and the dogs to roam. And no garden. I look at the bags and boxes of corn ears. My first and last corn harvest. Some farmers' daughter.

Are we wrong to leave, to give up so soon? If Dad could return to farming after it took his fingers, shouldn't we be able to stick out an unpleasant boss in order to stay on the ranch? Leaving was never an option for my parents; why is it an option for us?

I try to remember exactly when Dad harvested his last corn crop, exactly when he began saying goodbye to the farm. A blur of painful images flashes through my memory: the twice-daily whirr of Dad's breathing machine—his nebulizer—that helped to open his lungs after the initial diagnosis of emphysema; taking turns with Dad and my sister Diana to cook supper on the nights Mom was away at night classes, attempting to earn her elementary-education degree to help support us; the word "incurable" pounding through my brain when, as a high-school sophomore, I finally did a research paper on emphysema to learn more about the disease; winning an essay contest on the farm crisis, even though I didn't fully understand it; overhearing my sister Danette encourage Mom and Dad to move to the Black Hills, where she then lived, "if you end up losing the farm"; the pumping of the oxygen machine that invaded our hallway as Dad's disease progressed; overhearing their arguments about cigarettes, him saying, "How the hell else am I s'posed to deal with all this?"

"All this" was the double-whammy that struck in the early 1980s. Not only did Dad develop emphysema, which gradually stole his ability to work outside, but also, like so many others on the Great Plains, our family became deeply mired in the farm crisis.

For us, the crisis began showing its head when prices for farming essentials rose sharply in the late 1970s and early 1980s. I remember Dad once exclaiming, "What?" in a shocked, angry voice when the driver of the bulk-fuel truck told him the charge for filling our diesel tank. The cost was nearly double what it had been the year before, Dad said. And the price for seed, supplies, and equipment repairs increased, as well. Unfortunately, these higher production costs were not absorbed by higher selling prices for our crops. In 1980, for example, wheat cost $4.60 per bushel to produce but brought only $3.50 per bushel at market. A corn farmer, who spent $2.65 per bushel to grow his crop, received an average of $2.40 per bushel when he sold it. Extreme surpluses of corn and wheat— which some farmers blamed on government meddling with exports—only served to drive market prices lower. Those worst hit by the market glut were corn, wheat, and soybean farmers, and corn and wheat were Dad's primary farm crops.

The rising production costs and falling market prices ultimately led us straight into the credit crisis that so many others faced; our farm simply did not generate enough income to repay what had been borrowed against it. In the mid-1970s, when prices were good, Dad had bought a modest new Hesston tractor. Our family also moved into a new modular home in 1979 to replace the old farmhouse, which had become excessively expensive to heat and maintain. Dad borrowed money from the local bank and the Farmers Home Administration (FmHA) for these purchases. Like so many other farmers, he also borrowed operating money in the spring for fuel, seed, veterinary supplies, feed, repairs, and all other expenses; he expected to pay back this operating money when he sold cattle and crops in the fall. Unfortunately, the loan payments remained fixed, no matter how low market prices fell or how high operating expenses rose. Farmers Home Administration data shows that, in 1980, 27.2 percent of their borrowers were unable to make their loan payments; by 1982, nearly 34 percent could not repay. In 1984, the amount of money American farmers owed in interest payments alone was greater than their net farm income.

After a few years, a pattern developed: Dad would borrow in the spring, hoping a new year would be better; in the fall, when he sold, his income would not even cover what he owed his major creditors, let alone leave anything for our family to live on. I remember that Mom would get angry after Dad sold calves in the fall because the local FmHA office and the county bank often took every penny of the check. This meant we relied more and more on Mom's erratic pay from substitute teaching. She quit attending night classes when there was no money for tuition and fees. My sister Diana and I took jobs and assumed responsibility for buying most of our own clothes and supplies; Danette and David, our older siblings, were living on their own by now. Only Michele, the youngest, relied entirely on Mom and Dad. We all saved money and kept our grades up, too, knowing that our parents would not be able to help with college costs.

Once, on a summer afternoon after my junior year of high school, Dad approached me reluctantly: "Darcy, how much money do you have in the bank?"

"I don't know. . . . A little over $500, I guess."

"Could I borrow $150 to pay a bill? I'll pay you back when I can."

"Sure. Do you need more than that? You can have it all."

"No, that should be enough."

As I wrote the check, I said, "Dad you don't have to pay me back, you know." But the money was back in my account by the end of summer.

Of course, our family owed money to more than just the FmHA and the bank, and we didn't always have enough to pay them. Because we lived in a small town, this meant that our neighbors—the mechanic, the feed dealer, the parts storeowner—often lost money, as well. One stinging memory remains: not long after Dad's death, the owner of the parts store, a member of a family we'd attended church with all our lives, refused credit to my brother for a thirty-nine-dollar part "until you pay your dad's bill."

At the deepest part of the crisis, Dad faced two choices: selling out, or remaining on the farm but declaring bankruptcy. Selling out meant giving up the farm where he'd planted trees as a boy; where he'd started out in farming with his now-dead brother, Fritz; where he and Mom had raised us five kids and mourned five others. Selling out also meant starting over somehow, even as Dad neared sixty and was increasingly debilitated by the emphysema. And, selling out meant moving his family off the farm, leaving the destiny of this piece of land subject to the whims of a new owner. This last consequence was perhaps the hardest to face.

Although it would be a shattering blow to his pride—an acknowledgement that he was unable to pay his own debts—Dad began making plans to declare bankruptcy.

My harvesting complete, my thoughts return to the present and I gather the sorted bags of corn to go into the house while Shawn rides through the yard on Milly, the Quarter Horse mare he gave me for an engagement present. Both are lathered with sweat; they've just moved the ninety heifer calves that Bob bought two weeks ago onto different pasture. Next week, Shawn has orders to start fencing off certain pastures and to begin building a pheasant pen. Bob has decided he would like to make this ranch into a hunting retreat for his financial clients. Worse, he's begun rumbling about the conservation easement that protects this 1,200-acre ranch from subdivision. He's mentioned several times that, if he can get the easement lifted, he would like to sell off a 180-acre parcel of hay ground southeast of our house.

"Hey," I call to Shawn. "How did it go?"

"Great! Your mare sure is one hell of a horse. Can't even tell she's been nursing a foal all summer."

"Well, you trained her; she should be," I remind him. He is grinning as he folds his arms and leans over the saddle horn. Pure joy glows from his sun-browned face; I see that look on him only when he is horseback. How often will I see it now?

"How long 'til supper?" he asks.

"Oh, half an hour, maybe more."

"Good, that'll just give me time to brush out ol' Mil and give her some feed," he says, and turns to ride to the barn door. The question that began gnawing at me earlier this evening now cannot be ignored: Are we giving up too easily? Dad and Mom endured drought years, blizzards that killed most of the cattle, unrelenting money problems, accidents, even the specter of bankruptcy—all for the love of a lifestyle and a place. We're letting a difficult boss take away joys that can't easily be replaced: riding out to check cows, going for walks along the creek, gardening, watching the girls play safely in the backyard, listening to owls hoot as we fall asleep at night.

I try to console myself with thoughts of better income, more time off, more opportunities to travel and enjoy our hobbies—all things that ranch life lacks. But I'm surprised to realize I no longer desire these things, or at least that I want them less than I want to stay on the ranch. Dad faced bankruptcy because to leave his South Dakota farm would have been to lose his soul. Shawn's soul also needs grass, open spaces, and horses. And I'm learning that mine does, too.

Salvation for Dad's farm came from the federal government's Conservation Reserve Program, or CRP, which arrived in the nick of time. Essentially, the government paid farmers for putting crop-land into reserve; that is, for taking it out of production to allow native grasses and other ground cover to return. Land put into the program would not be farmed or grazed for ten years and could be hayed only in periods of extreme emergency.

The program was controversial, to be sure. I remember when the CRP was just beginning in the summer of 1985, how an aunt railed against the program, saying it was "sinful" not to use land for food production when so many people were starving. But my dad often said that the world's food "shortage" was more a result of governmental meddling, not farmers' and ranchers' inefficiency. We had all seen pictures of the corn and wheat crops of the early 1980s, piled on the ground because there was no more room in the grain elevators. For Dad and Mom, the program made sense in most ways: the one-time payment would cover the remaining debt on the place, and the cropland that had been used for over fifty years would get a needed rest. The downside of the program was that there would be no farm income for ten years; actually longer. Mom accepted her first full-time job, as a library aide at our local high school.

I remember vividly one evening in the summer of 1988, when Dad and Mom decided to enroll in the CRP. I came home past curfew that night, having been out late with a group of girlfriends, partying in a neighboring town. As I pulled into the driveway, I was giddy with excitement: in less than a month, I would leave for college in Montana. Montana! The place meant mountains, travel, glamor, and I couldn't believe I would be living there. I tried to sneak the screen door open, hoping that Dad would be in bed. Though I was eighteen years old and almost on my own, he still expected me to abide by house rules. As I eased into the entryway, I stopped short. There, bald head laid down on the kitchen table, Dad was asleep, waiting for me.

By habit, I stood perfectly still and watched to see if he was breathing. I constantly feared finding him dead. The doctors gave him maybe ten more years. Ten years. That's how long the land would be in the CRP. Suddenly, I realized that by signing up for that program, Dad was signing the end of his career. Even if he lived ten more years, he would not be able to leave the house to farm by then. At a time when my future looked so bright, his seemed so bleak; my earlier excitement evaporated. Finally, I saw his mutilated hand twitch, heard a ragged breath.

His nebulizer whistled emptily. "Dad," I called softly. "Dad, your treatment is over. You can go to bed."

He roused, looked sleepily at me, then at the clock. "You get home OK?"

"Yeah, Dad, I'm fine. Goodnight."

"'Night," he mumbled. He switched the hoses going into his nostrils, flipped on the oxygen machine in the hall, and shuffled off to bed.

"Shawn, can you clean off those lawn chairs and put them out in the backyard?" I call out the door. We're busy getting ready for our late-September harvest party, and already several of our friends have called to ask if they could bring along extra people. So, Shawn sets out more chairs as I make more salad and add corn to the pot. Laura, Carmen, and Maria get in the way more than they help, but all of us are enjoying the preparations. The busyness keeps our minds off more distant concerns: today, Shawn received a phone call notifying him that the construction crew won't need him.

Seems they wanted someone with more construction experience. Although we don't feel like we have any options now, other than waiting for the neighbor's lease to go through, I think we're both relieved that we don't have to move into town. I'm not sure what leaving the ranch life a second time would do to Shawn's spirit, or to mine.

The first time we quit ranching, he was afraid. He feared he couldn't do well at any other job—that he didn't know enough. "I've been ranching all my life," he'd say. Like Dad, Shawn was performing a man's work while he was still a boy: feeding out yearlings at age six, riding on three-day cattle drives by age eight, helping with calving by age ten. He never considered another way of life.

Now, though, he knows he can do other kinds of work. The question is whether he still wants to. When we left ranching in 1996, it was with the idea that other jobs would be better, would offer the kind of "successful" life people of our generation were leading. We returned because we finally acknowledged what our parents knew all along: that country life—ranching life—was the one we wanted. True success comes from living the life one is born to lead. Shawn has always known that he belonged on the land. It's taken me longer to realize, but as I settle into a lawn chair to watch the sunset and wait for our first guests, I can see that I do, too.

Another season, a different gathering. "Grandma, Santa's snoring by himself!" Laura races in to tell Mom that her mysterious Christmas decoration, a sleeping Santa doll that is supposed to snore only when someone turns it on, is snoring spontaneously again.

"I know, it just does that even when it's not turned on. I like to think that it's Papa, smiling down on us from heaven," Mom explains.

All twenty-one of us are gathered at the farm for Christmas this year; the group includes Mom, my siblings and me, our spouses, and ten grandchildren. Because Shawn and I usually prefer to have a quieter Christmas at our own home, this is only the second time we've traveled to South Dakota for the holiday since we married. The first time was in 1993, when Laura was barely six weeks old, the first Christmas after Dad died.

Although any weekend with twenty-one people crammed into

Mom's house is noisy and chaotic, being here this year is oddly comforting. I walk through the house with my girls, pointing out Christmas decorations that I used to help put up. We attend Mass at the church where we were all baptized, where our family has seen so many beginnings and endings. Mom cooks a traditional Christmas dinner; my brother, David, whose sideburns are beginning to turn gray and curl around his ears just like Dad's, sits at the head of the table. Later, he and Shawn go out to check on some of his cattle, which he pastures on the family land, land he is buying from Mom. Finally, we open gifts. The grandkids go first, taking turns from the youngest, my two-year-old, Maria, to the oldest, David's 14-year-old son, Justin. We laugh, tease, and sometimes cry over the presents; and Santa snores through it all.

Life flows in patterns around here, I reflect, as I straighten up from making holes for the girls to drop corn seeds into once again. One year ago we planted corn in a garden plot on a ranch we thought would be our home for years; now, we're breaking in a new plot, on a new ranch. Still, it is May, and we are planting seeds, just as I hope we will every spring, just as my dad and his dad did, just as my mom and her ancestors did.

Seven years ago this month, Dad died in a Veterans Administration hospital in a city he'd visited only twice, lying in a bed with machines all around, surrounded by all of us. We left Sioux Falls that sunny, surreal afternoon and drove the three hundred miles back to the farm to prepare to bury him. And then the rains came.

For the next several days, spring showers soaked and bathed the ripe earth. On the day of his funeral, while waiting for the Mass to begin, David and I gazed out at the drizzle through two tiny windows in the church doors. Someone nearby worried about how muddy the cemetery would be. "Don't cuss the rain," David said softly. "Look at that! Dad always loved to watch a good rain."

Today, I use the break in this spring's rain to plant corn. Shawn tilled up the new garden last weekend, and I've spent several days raking out the clumps and smoothing the soil. "Put the seed in the hole, Maria. See, just like Laura and Carmen are doing. There! OK, now cover it up." I repeat the instructions often, in no hurry to finish.

In February, we moved. The lease that our neighbor waited

on finally opened up, and Shawn assumed management of this twenty-three thousand-acre ranch in March. We now live ten miles north of Bob's place, farther up the valley, in a spacious four-bedroom home.

Life isn't perfect; Dad's life showed me there are no storybook endings. Shawn's hours are still long, and sometimes he works all through the weekends. We still fight about balancing the job with our growing family. But Shawn's opinions and experience are more valued here, and care of the land and the animals is truly first priority, put before making impressions on clients or neighbors. On this ranch, we are learning the techniques of Holistic Resource Management, a land-use theory that emphasizes healing the land, allowing it to repair itself. The actual techniques resonate back to what our parents taught us as we were growing up. It feels like our work here has a noble purpose: raising food while still caring for the land. Best yet, Shawn spends much of his time horseback.

I can't deny that Shawn is happy, so I'm learning to accept that the long hours are simply a part of ranching, and I try not to take it personally. I understand that no man's love for me should be measured in minutes and hours.

As I poke the last few holes in the ground, I reflect that I've come to this life because of the men I love. I'm a farmer's daughter because my dad chose to farm, and I learned his ways in spite of myself; I'm a rancher's wife because my husband chooses to ranch, and I can't ask him to do anything else. Like my mom and grandmothers, I've decided to honor the part of my identity that complements the men I love.

I stand and look at our new place: the big gray house, the corrals down the hill, the horse pasture out back, the tall cottonwoods where the wild turkeys roost. Our eighth home in nine years. But I'm learning that home is as much about a way of life as it is about place.

And I feel like I've finally arrived.

Crazy Woman Canyon beckons us this sultry, late-August afternoon with promises of cooler air held tight against her walls, nerve-soothing water tumbling down her creek, chokecherries ripened black on the bushes that line her road.

"OK, girls, grab your berry buckets; let's walk down this way a bit," I say as I unfasten car seats and help short legs reach the ground. I'm hoping we can pick enough ripe berries today to make jelly for the winter. Last weekend, most of the chokecherries were wine-red, not quite ripe enough, at least according to my field guide on wild berries. My husband Shawn thought they were ready, seemed to remember picking them that way as a kid in the Tobacco Root Mountains of Montana. But the berries pulled hard from the vine, and when I got the few of them we picked back to my kitchen, they hardly juiced at all, even after being cooked down.

This is my first summer picking and preserving wild berries. For years I've heard people talk about making chokecherry jelly and wine, but I've never learned how. The only instructions I have are cookbook recipes and Shawn's vague recollections from his childhood. So before we left the house today, I called my mom to see if she knew anything about picking and rendering chokecherries. "No, I never did that," she said. "They say it's so much work to extract the juice. But I remember your Grandma Lipp doing it."

Grandma Lipp, my dad's mom. Barely a memory of her exists in my mind. She stood less than five feet tall, with blue-gray hair and black-rimmed eyeglasses. She always wore flowered housedresses, often an apron, never pants. In her kitchen was a cupboard drawer—top, left-hand side—filled with candies to give to us grandkids during our weekly visits. She died when I was seven.

"Look, here's a bush just full of berries," I call back over my shoulder to Shawn, who's walking slowly with Maria, our youngest daughter.

"Are they ripe?"

"Looks like it—they're pretty black. The birds have eaten most of the ones on top." I begin pulling a few berries at a time from the bush, dropping them into the ice-cream bucket that dangles from my arm.

"Here, do it this way," Shawn suggests, and shows me how to hold the bucket under a cluster of berries, then strip away the whole cluster in one motion, like milking a cow. Last week, the berries weren't ripe enough to strip this way. *Plunk-plunk-plunk*: they sound like marbles hitting the bottom of the bucket.

We spread out up and down the lane, the girls picking lower branches, Shawn and I the upper. The canyon's echoes warn us of oncoming vehicles, and we periodically grab hands and buckets to duck closer to the thickets. Laura, who is almost six, frequently stops to eat a berry, spitting out the poisonous pits. I wrinkle my nose at her.

"What? I like them," she says, mouth wide in a purple, front-toothless grin.

"Ugh! How can you?" I exclaim. "They're so sour!"

"Mom, will the jelly we make taste sour, too?" our four-year-old, Carmen, worries. Carmen dislikes anything sour, even lemonade.

"No, we'll use sugar to make the jelly. It will taste good. Just think—if we pick enough, we can have chokecherry jelly all winter, and every time we eat it, we'll remember berry picking!"

"I like berry picking, Mama," Laura says.

Grandma Lipp didn't live long enough to introduce me to berrying; it was another lady, a grandma very much like her, who sparked my interest in gathering the wild fruits in the mountains near our home. Grace babysat our girls for three years when we lived in another part of the state. In fact, our daughters loved her so much they began referring to her as one of their grandmothers.

Providing food for her own family, as well as her daycare children, occupied much of Grace's time and attention. In March, the sunlit counters of the daycare's playroom filled with trays of potting soil, tiny seeds hidden beneath the surface. By late spring, the children's preschool projects squabbled with adolescent seedlings for space; I've seen tomato plants growing there that needed staking before they ever went into the garden outside. In summer, the older kids accompanied Grace to the garden and greenhouse. All fall, her food dehydrator hummed and her kitchen stove simmered with the task of preserving all that bounty. "Grace, you do so much," I'd say, awed by her productivity.

"Oh, well, you know, I require very little sleep," she would always say, waving off the hours of work. "And besides, we do have fun!" Her face and shoulders would scrunch up in a whole-body grin. Her energy brought back faint recollections of my Grandma Lipp, who never seemed to age or tire.

Late one summer, just before Labor Day, I asked Grace if she and her family had plans for the weekend. "Well, yes," she said. "Every Labor Day, we go up to Union Pass to pick the wild currants there."

"Currants?"

"Oh, yes. They're little red berries, about as big as my fingernail. The whole pass is just covered with them! I take the grandkids and we pick and pick—we fill whole five-gallon buckets full! Oh, and they do make the best jelly!"

That weekend, I casually suggested a drive to the Lake of the Woods, on top of Union Pass. Sure enough, as we picnicked and explored the lakeshore, I did find occasional red berries like those Grace had described. I picked a handful and asked Shawn if he knew what kind of berry they were, but he was also unsure. We decided not to pick any.

Leaving those berries there, though I'm sure they provided ample food for the birds and wildlife, still didn't sit well with me. Some part of my womanly nature needed to preserve that treasure, to use that bounty somehow to fortify my family. I could not walk past those tiny bushes without feeling awed at so much nutrition and goodness, all free for the picking. So, I decided to continue my berry explorations and bought a field guide. The next summer, in a different mountain range, we headed up Crazy Woman Canyon, looking for wild raspberries. It was late July, prime season for them. We started looking in a spot where we'd seen blossoming bushes during our Fourth of July camping trip earlier that summer.

The canyon didn't reveal many of her wild raspberries that day. She did, however, uncover another treasure—mountain gooseberries—plump, purple globes hanging over the creek banks on thorny vines. Picking them, even gloved, pricked our fingers, so we let the girls play while we sacrificed our calloused, grownup hands. But they tasted too wonderful not to pick—bursts of tangy-sweet juice exploded from each bite. I imagined gooseberry jelly on fresh rolls with venison steak and fried apples in the fall.

When I got the berries home, I discovered that the picking wasn't the only hard part. Although my mom had taught me to make jam from the whole fruit, I'd never attempted to extract just a fruit's juice for jelly. I didn't have the necessary equipment, so I spent hours one morning pressing cooked berries against a sieve with the back of a wooden spoon. Then I stood at the stove stirring the juice and sugar together until the natural pectin in the berries caused the mixture to thicken. At the end of the morning I had five jars of deep purple jelly. I felt pride at making the treat with only the equipment I had on hand.

Grandma Lipp must have known how to make do, as well. Family stories abound with details of how she used whatever resources she had. In our family photograph album are pictures of my older sister and brother sporting the matching outfits that Grandma Lipp made for them from the fabric of men's torn work shirts. When she was still able to drive out to the farm from town, she harvested the horseradish plant that grew in our farmyard, peeling the tough stalks and grinding them by hand with vinegar and spices to make homemade sauce. Grandma processed all the meat from the pigs Dad and his brothers butchered, showed my mom how to fry the testicles from castrated bull calves at branding time, and asked Mom to save the chicken feet from our late-summer butcherings so she could pickle them. While my maternal grandmother embodied domesticity laced with a heavy dose of propriety, to me Grandma Lipp was domesticity mixed with grittiness, wildness, rawness. As I now clamber up a slope to pick the top of a chokecherry bush, my hands are already stained violet and my fingernails are ragged and dirty; I smile, feeling a kinship with my dad's mom.

I didn't feel the same way ten or fifteen years ago. Then, I felt sorry for my grandmother, for the lack of opportunities she had. I thought then that a life revolving around domestic chores was boring and unfulfilling. I promised myself I would never be "just a housewife," like she was. When I filled out college applications, I always neglected to mention my numerous 4-H awards, even the "Top Home Economics" award that earned me a trip to Denver, Colorado, in 1985. Away at school, I ate all meals in the campus food center or survived on ramen noodles and instant macaroni and cheese. When friends baked special birthday cakes, I pretended I didn't know how to help them, even though I'd won awards for cakes I baked for the county fair at home. Domestic skills were so embedded in my childhood, in the heritage I had received from my grandmothers, that they felt like a trap. If I fell into them, would I ever be able to climb out into making a life of my own?

I changed out of necessity. Moving into my own apartment meant re-teaching myself how to cook, and out came the boxed-up cookbooks my mom had given me. I began to feel a little pleasure in spending an evening after classes baking brownies or cookies to share with my neighbor across the hall. When I met and married Shawn, we adopted traditional husband and wife roles because my school schedule allowed me to be home early enough to start supper. I experimented with countless recipes, serving anything from rancher's round steak to Chinese food. Cooking and keeping house became sources of pride, much to my chagrin.

And now, with three daughters to feed, my domestic chores have multiplied. We are here today, spending our Sunday afternoon picking berries, because I like to serve my family food that I have prepared myself; food that is prepared with love and care, not chemicals. The irony sometimes shocks me: after quitting my teaching career to raise my children, I have become "just a housewife," like my grandmothers. The difference is that I made my choice freely and consciously; cultural norms did not decide for me the way they did for my ancestors. On the contrary, modern society often makes me feel ashamed of my choice, as though I live a diminished life by being "just" a homemaker instead of a teacher, CEO, or pediatrician. Although I strive toward a writing career, I find my life's real work in nurturing my family, in doing "women's work." I wonder if Grandma Lipp, who did not have a choice, still felt this satisfaction.

I long to ask her, to talk with her now, as an adult woman; the seven years I shared with her are a scarce source of her wisdom. I have Grandma Lipp's candy dish, her height, her poor eyesight.

Other parts of her legacy are lost on me. I sew, but only with a pattern to follow. I cook, but not from her recipes. I know how to make my maternal grandma's roast beef and my mom's fried chicken, but I don't know how to make brats and sauerkraut like Grandma Lipp. I'm more likely to cook spaghetti or tamales for supper than anything from my German heritage. And although I've baked hundreds of chocolate-chip cookies, apple pies, and white velvet cakes, I still don't know how to make *blachenda*, the pumpkin-filled turnovers my dad loved from his childhood.

Some things I do know: that Grandma was born in North Dakota, the daughter of German immigrants from Russia. That she married my grandpa, himself an immigrant, in 1920, at the age of eighteen. They started their life together farming on the border between the Dakotas, near a tiny Indian reservation town. She bore nine children there, including Dad. In 1944, with her three oldest sons at war, she moved with the rest of the family to a farm near Glencross, South Dakota, settling among German-Russian neighbors and making a home there for twenty years. Just years after her oldest sons returned safely from war, she watched her other two go again, this time to Korea. She never lost a son in war, but she mourned two of her children—Fritz, who died moving cattle in a lightning storm in 1967, and Betty, who died of lupus in 1968. At the end of their career, she and Grandpa left the farm, retiring to a tiny purple house in town so that we, the grandkids, could grow up in the country. In 1977, Grandma died of emphysema.

Skeletal facts. A biography, nothing more. I want to know so much. How did she endure the endless hours alone, raising all those children while Grandpa worked the fields? Why did she raise her daughters cloistered in domesticity while only the sons worked in the fields? When did she decide her home would have a windbreak hemmed with nearly a quarter mile of lilac bushes? And how did she persuade my practical grandfather to plant them? How did she feel, sending all five of her sons off to wars? Who comforted her through the loss of two adult children? How did she keep faith in her God through all that? And why, having battled asthma and emphysema for nearly forty years, did she finally have to succumb to it?

I crawl back down the hill, returning to my family at the bottom of Crazy Woman Canyon. The girls are begging to cool off down by the creek, where snowmelt tumbles over rock. Local legend tells that a ghost—a crazy woman—haunts this canyon. According to the story, the Crow Indians allegedly killed a white man and his children in the canyon, although the motive for doing so is unclear. The Crows inhabited these mountains, the Big Horns—the Shining Mountains, to them—and resented and resisted white incursion. The story says that, upon finding her husband and children murdered, the woman went crazy, and that her cries still echo here by the creek at night.

Her presence doesn't frighten me; if she is here, she's simply looking for her family. I understand how such a loss could bring about insanity as I look at my three sleeping daughters nestled asleep in the back seat of the pickup, sunburned and dirty, but safe. In one horrible afternoon, probably in less than ten minutes, the woman lost not only those she loved most, but her entire purpose for living. I daydream for a moment, imagining that she was spared from the slaughter because she was down at the creek hauling water, or perhaps even gathering berries—doing some chore that was part of caring for her family. The story says that the Crows allowed her to continue living in their territory because she had been "touched"; I say that the Crows knew killing her would be a mercy. As we drive down to the mouth of the canyon, I silently thank her for allowing us to enjoy the canyon's bounty and pray for her soul.

Grandma Lipp was "touched" by loss, too: loss of homes, of crops, of children. How did she keep going? Even as my mind asks the question, I begin to know the answer from deep in my woman's soul: she went on because there were others to care for, because there was still work to be done. Like our friend Grace, my grandma stayed youthful because of the children and grandchildren who relied on her. The woman in the canyon lost her mind because she had no more reason for living. And in this instant, I knew that I will not be ready to leave this world until my own real work is done: my own children raised, my own grandchildren growing. Grandma Lipp let the emphysema take her because, at age seventy-seven, she was finally ready to rest, her life's work complete.

On the day after our chokecherrying, I stand once again at the kitchen sink, washing and sorting berries. I seem to know instinctively which berries will juice. After testing a few, I discard those that don't have the right feel. I am amazed at how much easier today's jelly making is; it's as though I'm being guided. The connection is there, though fragile. Perhaps this autumn, after the pumpkins ripen, I'll find a recipe for *blachenda* in my Catholic women's cookbook from back home. Maybe I'll even serve it with brats and sauerkraut.

By noon, I have twenty-three jars of burgundy jelly lined up on my counter. Along with the jars of gooseberry jelly I made last week, it is enough for the winter, the only expense having been sugar, time, and sweat. I think Grandma would be proud of me.

We have chokecherry jelly on toast the next morning, after it has set up. The spread tastes sweetly piquant, like store-bought cherries mixed with sage and mountain air. "Mommy, this is good! I want some more!" the girls exclaim. As I reach for more bread, my imagination hears the mournful wailing of wind through Crazy Woman Canyon, sees the selfless grin of the Union Pass mountain ridges, and feels the welcome cool of a chokecherry draw pocketed into the South Dakota prairie.

Friday morning . . . cool, cloudy, the scent of rain hanging in the damp air. I ready Laura and Carmen for school. For an hour, the house flurries with activity: cereal to pour, hair to braid, backpacks to check for books and mittens. Then, hoods up against the chilly air, the girls walk down our driveway to the bus stop. I watch until they're safely on board. The house is silent now, our youngest daughter, Maria, downstairs playing with dolls.

I drain the last of my coffee, pour a fresh cup, then reach under the counter for my big stainless-steel bread bowl. A good day to bake bread. Fall is finally here, when baking doesn't seem like an impossibly hot chore, but a pleasurable way to warm the kitchen. Normally, I buy a month's supply of soft wheat bread from the store. But when I can, when the weather's cool and the garden's been harvested, I like to bake my own.

As I take the yeast from the refrigerator, I think about what kind of bread to bake today. I don't make the standard white loaves my mom did—I'm not baking for efficiency or economy, but for quality. Sometimes, chewy oatmeal loaves line my counter after a baking, but today I'm in the mood for hard, hearty whole-wheat.

Grandpa Lipp, my dad's father, farmed wheat in western South Dakota and passed the craft down to my dad. I knew my grandpa only in his retirement, but his gruff, no-nonsense manner told me even then what type of farmer he had been. Grandpa stood only a little over five feet tall, with severe black-rimmed glasses and a face that only rarely smiled. Although he would tease us kids and enjoy watching us play, he expected to be obeyed without question; he was not an approachable grandfather, not the kind whose lap you would clamber onto for a story and a tickle. Rarely did he discuss anything other than crops and weather when we visited on Sunday afternoons. When he came for extended stays on our farm, after Grandma Lipp had died and he had moved to a veterans home, he would still sit for hours on our front deck, surveying the fields. As he grew older, he spoke his native German more often, preserving for himself a heritage—of a people, a language, a way of raising food—that was dying. Although I reconnect to my agricultural roots by growing corn, a plant I can manage in the space of my garden, wheat is truly the crop of my people.

Grandpa's family formed part of a little-known immigrant group: Germans who for generations had lived and farmed the steppe of South Russia, now the Ukraine. These Germans from Russia, sometimes called German-Russians, dominated the population of the north-central area of South Dakota where I grew up. Indeed, an appendix of early immigrants listed in author George Rath's history, *The Black Sea Germans in the Dakotas*, reads like the telephone directory of my hometown: Aberle, Bader, Bollinger, Engelhardt, Feist, Fischer, Jung, Heib, Holzer, Keller, Kempf, Kraft, Kramer, Lippert, Meier, Pfeifle, Reich, Richter, Schumacher, Selzer, Welder.

These German families first moved to Russia in 1773, when Empress Catherine II, also called Catherine the Great, encouraged immigration to her country, particularly to the empty steppes on the Volga River near Saint Petersburg. She offered settlers religious freedom, plus freedom from taxes and military conscription, in return for development—meaning economic utilization—of the land. Farmers from several countries, but especially from crowded Germany, accepted her offer. In 1804, pleased with the progress the German colonists were making, Emperor Alexander I promoted further colonization, this time in the Black Sea region near Odessa. He enlarged his grandmother Catherine's promises by offering thirty to sixty hectares of land to each settling family. However, Alexander accepted only those immigrants who were fathers and property owners in Germany. German farm families flocked to the steppe and raised products suited to the land and its climate: cattle, sheep, horses, and wheat. The ancestors of both my paternal grandfather and grandmother were among those who settled near Odessa, as were most of the ancestors of the Germans from Russia who finally emigrated to North and South Dakota.

The colonization of the Black Sea region was even more successful than the colonization of the Volga. By 1837, over fifty thousand colonists settled there, adding to their original hectares by purchasing land with funds collected from the colonies for that purpose. The colonists soon owned nearly a quarter of the arable land there.

Such success caused envy among Russian nationalists, and when Alexander II assumed the throne in 1855, the tide of favor for the Germans began to turn. Upper-class Russians, in particular, disliked them, and proponents of a Panslavistic movement urged authorities to force the Germans to become Russian in both religion and culture. Alexander II revoked the many of the colonists' earlier privileges beginning in the 1870s. The first groups of emigrants left Russia shortly after these changes, including some who settled in the Yankton area of southern Dakota Territory. Decrees by Alexander III in 1890 and Nicholas II in 1897 established Russian as the only language allowed in schools and public offices, even those serving the German colonists. The loss of the freedom to speak their native tongue, coupled with the reinstatement of military conscription and a shortage of suitable land to begin new farms, led in 1894 to another wave of emigration headed for the United States. This one carried my grandpa, then three years old.

As most ethnic groups do, the Germans from Russia settled in enclaves of their own kind in North America. These people, however, bypassed the large cities in the east to head straight for the Great Plains of North and South Dakota and the prairie provinces of Canada, where they'd heard the land closely resembled their beloved steppes. There, they could continue the livelihood of their ancestors—growing wheat.

I dissolve the yeast in warm water, add eggs and shortening, and then finally begin mixing in cupful after cupful of stone-ground wheat flour. Already the bready scent of warm yeast and flour sparks a pang of hunger. Soon the dough is too stiff to stir with my wooden spoon, and I squeeze in the last of the flour by hand. Then I dump the dough onto the counter to knead. I fold over an edge of dough, roll the mass away from me, give the mound a quarter turn, and repeat. Fold back, roll out, turn. Fold, roll, turn. Kneading requires ten minutes of shoulder-aching work, but I lose myself in the rhythm of the folding and rolling, in the silky feel of the dough, in the smell, so that ten minutes flows by without me really noticing.

Today, for less than one hundred dollars, you can buy a bread machine to do most of this process for you. A bread baker simply measures ingredients, sets a cycle, and walks away. In a few hours, a fresh, warm loaf awaits. But I don't have a bread machine. I could buy one if I wanted to. I don't.

As I knead, I reflect that Grandpa would approve of the way I'm spending my morning today more than he would most of my mornings. True to Old World traditions even until the end, Grandpa believed women should not work the fields. My three aunts—his daughters—tended the house and garden with my grandmother, but not the wheat crops. Even the next generation followed this division of gender, at least in my family. My dad, father of four daughters and only one son, still did not allow my sisters and I to help with the farm work. I finally learned to drive a tractor at age seventeen, and then only because I worked for the city maintenance department in my hometown.

In addition to believing women should not do farm and ranch work, Grandpa believed women should not work outside the home or pursue creative expression beyond handicrafts. His way of thinking, echoed by patriarchs of farm families throughout much of the Dakotas, perpetuated the traditional family structure even when change was sweeping through other parts of the world. When a generation of women began expressing feminist ideas and leaving home to work in the early 1970s, my mom, along with the moms of my friends, was still at home baking bread.

Since Grandpa Lipp and his family were some of the last German-Russian immigrants to reach the Dakotas, they searched long for land to homestead. I imagine them traveling from Chicago to Yankton, South Dakota, and traveling north by wagon through South Dakota and over the border into North Dakota. Earlier settlers had already claimed most of the flat, black-soiled land of the eastern Dakotas. Grandpa's family made their home among a Catholic German-Russian enclave near what became the town of Hague, in south-central North Dakota, just a few miles east of the Missouri River.

After marrying, Grandpa again searched for land. He and my grandma first farmed near a tiny town on the Standing Rock Indian Reservation, just south of the state line and across the Missouri

River to the west. Then, they delved a bit further into West River country—as the land west of the Missouri is commonly known in South Dakota—settling on the farm near Glencross, South Dakota, where my family still lives. The land here, though flat prairie in places, also undulates with the sort of draws and coulees that mark the foothills of the Rockies. Farming and tending this land would require more attention, more care.

In farming, Grandpa's Old World upbringing served him well. Like most of the farmers in the early part of the twentieth century, Grandpa relied on what agronomists now call "diversification" to keep his farm, and his family, alive. The parcels of land that were relatively flat were put into crops, as this type of terrain could be tilled with minimal erosion of the rich topsoil. Hilly, rough land was better suited to pasturing livestock, who then converted the native prairie grasses into human food through their meat. On their first place, he and Grandma raised crops and livestock and ran a small dairy operation—as close to self-sufficient as a South Dakota farm family could become. On his second farm, Grandpa milked cows, raised hogs and beef, and grew corn, oats, and wheat—always wheat, maintaining that same self-sufficiency and diversity.

In the early years of our marriage, Shawn and I lived in north-central Montana, in the middle of the largest wheat-producing region of the United States. Shawn had lived and graduated from high school there, and we moved back two years after our wedding to manage a small cattle herd for a large grain farmer. The town in which we lived nestled down into a valley next to the Missouri River. When I would drive up onto the flats above town, I saw only field after field of wheat, punctuated by farmhouses, machinery sheds holding expensive tractors, and grain silos. A visit to my in-laws, who lived thirty miles out of town, left me reeling from a kind of vertigo. Miles and miles of tabletop prairie glistened gold with wheat, the furrows running in endless parallel lines across the dips and swells of the landscape. A friend from Montana referred to this region as the "wheat desert," and it didn't take much imagination to see why: if I squinted just right, the gold of the wheat became the tan of sand dunes, covering any signs of life or diversity.

During my time there, I learned that some of the wheat fields were so large that the furrows themselves were often several miles long. Every farm was huge, at least a couple thousand acres, and the economic pressure to get bigger led to tearing up more prairie, pasturing cattle on leased ground hundreds of miles away, buying more expensive equipment. Those farmers who could successfully manage to grow kept enlarging, while debt and dwindling production rates swallowed the others.

We knew farmers there who struggled to survive on small, manageable farms. However, the "green revolution" that swept through farmlands in the 1970s continued to push them to enlarge. With advances in chemical fertilizers and pesticides, proponents of this revolution planned to conquer world hunger. Popular thinking dictated that if a thousand acres was good, then two thousand must be better. The small family farmer was forced to search for more efficient methods to increase production and to economize, not by remaining small, but by focusing all energies on one or two products—usually a small herd of cattle and millions of bushels of wheat.

The discomfort I felt living there was only partially based on the immensity of those wheat fields. Along the drive to my in-laws' house, a set of white cylindrical tanks sat at the top of a rise, standing a full story tall. I learned these tanks held anhydrous ammonia, a chemical fertilizer applied regularly to the land to infuse the crops with nitrogen. Indeed, I had seen tractors pulling contraptions that looked like tanks of water. When I finally asked what type of implement that was—a bit embarrassed that I didn't already know—Shawn explained that those were the fertilizer tanks, dumping liquid nutrients onto the ground.

By then I had lived twenty-five years on and around a farm, yet I'd never seen land fertilized with a chemical. To me, fertilizer was the manure my dad put on the fields, using an old blue manure-spreader. He shoveled or used a tractor loader to fill the wagon with cow or pig shit, then drove through the field while the dried waste sifted down onto the ground through a paddlewheel at the back of the wagon. The manure, really recycled grass and alfalfa, still contained ample nutrition, since animals absorb only about one-third of the nitrogen in the plants they consume. In addition, the dried plant material encouraged microbial activity to rebuild the topsoil; a slow, glacially-paced process needed to counteract the erosion of topsoil by wind, water, and western agriculture. Dad and Grandpa couldn't fertilize every field every year, but they tried to enrich each field at least every three or four years. Grandpa also taught Dad to leave the wheat stubble—the short stalks remaining after they harvested the grain and straw—standing in the field. By early spring, heavy, wet snows rotted the stubble, and this rotted plant material was then plowed back into the ground—Mother Nature's own compost. Often, Dad would graze the stubble fields in the winter, thus bypassing the step of collecting manure to spread on the fields; the cattle simply pooped where they walked, leaving fertilizer behind them.

The big farms of Montana left the stubble in the field as well; it not only enriches the soil but holds it against the omnipresent wind. But fertilizing with organic material was simply impractical for farms that enormous: the small cattle herds most farms kept couldn't produce enough manure to fertilize all the fields in even ten years' time.

Along with diversification, Grandpa's model of land stewardship included crop rotation. For example, after several years of growing a nitrogen-draining crop, such as corn, in a field, he would either let it lay fallow or reseed it to wheat, oats, or alfalfa—all crops that don't require such heavy amounts of nitrogen. The soil then had time to regenerate, and the pests that bother particular crops would die off. Studies show that modern farms that raise only one crop—a practice called monoculture—rely heavily on chemical fertilizers and pesticides to do what crop rotation did for Grandpa.

It is not too far a stretch of the imagination to see the land as female and to realize that the way industrial farming treats the land is much like sexual exploitation. Chemicals are applied to regulate, even force, the land's fertility. Land that is not meant to be plowed, land that is better suited to raising grass and livestock, is still torn up, "raped" for the purpose of creating larger farms. In a desperate cycle, landowners feel forced to exploit what they do have—the land—to get what they don't have—money. Although I don't much agree with Grandpa's idea that women needed protection from the harsh life lived on the farm, I do agree that the land, our eternal mother, needs and deserves protection from the hard, money-hungry industrialism that dictates farming today.

I place the kneaded dough back in the bowl to rise, following the process my own mom taught me many years ago, a process she learned from her mother and grandmother. Irony makes me smile: although women have for centuries been relegated to the role of preparing food from the wheat after the harvest, archeological evidence suggests that the first wheat farmers were probably women. It likely would have been women who noticed that seeds of wild grasses, often chewed for food, germinated and produced more grasses the following season. It would have been women who experimented with planting a few seeds near a shelter to see what happened. And it would have been women who tended a crop while male members of a tribe were out hunting.

The women who had been Shawn's friends during high school, and who later became mine when we moved back to his hometown, did help to tend the wheat crops. I suppose the sheer size of those farms overruled any traditions about female farmers: all available hands worked in the wheat fields, regardless of sex. One good friend spoke affectionately of "her" Big Bud tractor—a dual-wheeled, twenty-foot-tall behemoth—like I might speak of my horse or dog. However, many of these young women still left the fields shortly before noon and in the early evening to help their mothers prepare meals, especially when the large harvest crews moved through the area in late summer. After serving the meal and cleaning up, the women returned to the fields to work with the crews. Gender equity, perhaps, was just as elusive on the high plains of Montana as it was on the prairies of the Dakotas.

The bread dough now fills the bowl almost to the rim; an hour earlier, the mound was half this size. Time for my favorite part. I set the bowl back on my work counter, double up both fists, and punch down. Trapped air escapes with a soft hiss, and the expanded dough deflates slowly. With my knife, I cut the dough into thirds to make three loaves. Again, I've not chosen to be as efficient as traditional farm and ranch wives. My mother-in-law often speaks about baking ten to twelve loaves twice a week when she was cooking for the men who worked with her husband. Maybe it's no wonder that she now uses a bread machine.

By contrast, it's the process of bread-baking that I enjoy, perhaps even more than the product. Mixing, kneading, rising, shaping, rising again: like a small child engrossed in a game, I delight in each step of play. Having a machine do the work for me would still give me the homemade bread I want to feed my family but would take away most of the fun of creating it.

Some would say the changes in wheat farming from my grandpa's small farm to the large-scale industrial farming of today are but changes in the process, designed to produce the same product more efficiently, more economically. Indeed, neither Grandpa nor Dad grew rich farming and at times struggled to survive. Their methods—diversification of crops, maintaining a farm size that one family could manage, relying on nature rather than chemicals—had become archaic. By the time Grandpa died and Dad's health forced him to retire, the tide in farming had turned away from the methods of the immigrants and toward industrialism.

Rare now are the family farms that operate without hired help, or that do not rely on chemicals, huge equipment, and more acreage to make a living. Rare are the farms that diversify. Common now are those that focus on one crop, and possibly one type of livestock, per year. Ironically, if isolated from the commercial food supply for more than a few months, the families of these large farmers, the same families who are to produce food to feed the world, would likely suffer malnutrition and starvation almost as quickly as non-farming populations.

The bread loaves are ready to go into the oven now. I wonder whether bread machines produce the same incredible aroma as oven-baked bread. As that heady scent fills the house, I anticipate my snack: the heel of the loaf, cut while still steaming so the butter melts instantly.

But along with my anticipation, I feel sadness. I realize that this morning's baking is simply a luxury, a hobby of sorts; I can easily drive to the store for my bread. In other countries, children like my Maria don't have even a crust for their meals. In centuries past, the lack of flour to produce bread sparked revolutions. Landowners striving to increase the size of their farms, in order to export their wheat to richer countries, pushed peasants into the cities, where children starved because families no longer had access to basic ingredients. Marie Antoinette's sarcastic, "Let them eat cake," spoken at the beginning of the French Revolution, underlines the disregard of rich landowners for both the land they owned and the poor they were supposed to feed. In Russia, not long after my grandpa's family left, this peasantry revolted against the royalty and all those associated with them, including the German colonists. Some of the last waves of Germans from Russia who came to the United States in the years prior to the Russian Revolution in 1917 never again heard from the family members left behind.

I am also sad because I know that although I bake the bread myself, I am still using wheat flour from the commercial supply, a supply grown in ways so different from the way Grandpa raised wheat. I know that simply by eating from this supply, my family and I are perpetuating the cycle of chemical- and petroleum-based industrial agriculture. Unfortunately, the "green" revolution has done little to solve either world hunger or environmental concerns. The price of a loaf of bread hovers near three dollars; soil erosion continues at a pace hundreds of times faster than that in which the land can regenerate; and small, family-run farms disappear every day. Farming is no longer a lifestyle as much as a big business. I am sad for the end of a way of life and grateful that neither Grandpa nor Dad lived to see it fully.

I'm sure they knew it was coming, though, the same way those original German settlers probably suspected the Russians wouldn't welcome them forever. I think of the one picture I have of Grandpa Lipp. It's a family picture, and in it he stands at the center, with Grandma, surrounded by their nine children. Although he lacks height, he makes a formidable figure there, in his stiff-looking suit, mouth unsmiling, eyes staring straight at the camera. It's as though he can see change coming—has lived through so much change already—and is facing it down. I imagine him thinking that if he just sticks to what he knows is right and teaches his children the same, he himself could stop that tide from turning.

I wish he could have.

Dark gray thunderheads boil out of the northern sky as my family and I climb to our seats at the Sheridan County rodeo arena. Our tickets tell us we're to have seats 10 through 14, but when we find our spot on the narrow metal bench, we can squeeze only into seats 11 to 13. On both sides of us, eager rodeo fans fill the bleachers. The humidity makes us feel sticky and sweaty, and my daughters complain of the heat.

Soon, though, the action begins in the arena below us. Besides viewing the typical rodeo events, at this show we will watch Indian horse relays. In these races, teams from Wyoming, South Dakota, Montana, and Idaho compete for top honors. Each team consists of one rider, who is dressed in tribal regalia; two "muggers," who hold the horses; and three horses painted like war ponies of long ago. The rider mounts the first horse, races around the track encircling the arena, dismounts, jumps onto the second horse, and continues on. The horses are ridden bareback, in the traditional way. The first rider to make three laps on three different horses wins the relay.

As we watch, my daughter Carmen notices that one rider sports a full headdress—probably Crow or Lakota. "Mama, are those real Indians?" she asks, brown eyes wide. I nod yes, hoping to shush her. Instead, she gasps and says, louder this time, "You mean we have *real Indians* around here?"

I feel my cheeks reddening as I respond, "Yes, we do. Indians are regular people." What I don't tell her is that, seated uncomfortably close to me, are two women in whose dark skin and hair, high cheekbones, and familiar accent I instantly recognize American Indian blood. All I can do is smile apologetically and hope they are not offended.

Ironically, if Carmen's great-grandpa Lipp had still been alive, he would probably refer to her as his "little Indian." Her dark eyes, tanned skin, and long brown braids resemble my little sister Diana's, and Grandpa called Diana his "little Indian" so often that I thought then that our family was American Indian. After all, my dad's skin was also dark, and he and my three sisters shared black hair. They looked like the people I saw around me growing up on the Cheyenne River Indian Reservation; I made the assumption easily.

Consequently, I felt hurt and confusion at the racial remarks I heard neighbors and relatives make: "Lazy as an Indian." "Be easy for us to have a new pickup, too, if we had the government handing us money." "Oh, he's not like the others; he's a *good* Indian." It wasn't until I realized my family's coloring resulted from German blood, not Indian, that I blithely accepted the racial prejudice that underscored much of life on the reservation.

Guiltily now, I listen to my six-year-old daughter express the same prejudice: that American Indians are the "other," a race to be mistrusted at the least, feared at the most. Although Carmen's perceptions grow from too many childhood tales of cowboys and Indians, the fact that I've never reversed her belief speaks volumes about my own prejudices. Indeed, I'm afraid of offending the women sitting next to me not only out of a sense of courtesy, but also because I still fear their heritage. Growing up, I knew several white girls who fought with Indian girls; none of the white girls ever won. I'm embarrassed that this thought even enters my mind nearly fifteen years later, but the reaction has been trained into me. Simply recognizing American Indians around me sends signals that prompt me to be more cautious.

I didn't always succumb to the prejudice. I remember being friends with several of the American Indian girls in my school, never thinking about any differences among us. Indeed, since my family survived on a small-farm income, I found being friends with the Indian girls much easier than being accepted by some of the rich white girls in my class. By high school, however, only one Indian girl remained in our class, and she was marked as different, if only by fault of being the minority. Still, despite these early friendships, two years ago I watched the Rapid City news with no real surprise when they aired a story about a former classmate—back then a shy, quiet girl who attended only the fifth grade with us.

Police had just arrested her for killing another woman in a fight over a man.

My family's farm lies within the boundaries of the historic "six-mile strip," a six-mile-wide, eighty-mile-long strip of land running from the rolling pastures of West River country to Mobridge, established as a railway station on the east bank of the Missouri River. To the north of the strip lay the Standing Rock Indian Reservation; to the south, the Cheyenne River Indian Reservation. The strip was originally designated to serve the cattle industry; cowboys could drive their herds to the rail head without crossing tribal land. All along the strip, towns sprang up at roughly ten-mile intervals—the distance of an average day's drive, or about as far as cattle will travel from watering hole to watering hole. These towns—Glad Valley, Isabel, Firesteel, Timber Lake, Glencross, and Trail City—in turn attracted settlers, primarily German-Russians and Scandinavians. Because the strip was designated nontribal land, these settlers could homestead here. Later, when cowboys no longer used the strip to trail cattle, the white settlers remained.

So, besides growing up with strong ethnic traditions from my European ancestors, I interacted daily with Lakotas from the two reservations. Ironically, I learned few of their customs. The current interest among whites in American Indian dress, dance, and spirituality is relatively new. While I was growing up, I did not associate my Indian classmates with powwows, the Ghost Dance, and the return of the wolf, bear, and buffalo. Instead, I learned to associate them with the empty beer bottles, run-down houses, and packs of stray dogs I saw whenever we drove into the government housing development south of town. A harsh reality, but the truth, nonetheless. My ancestors no more romanticized the Indians than the Indians glorified the whites. Living side by side erases many of the prettier myths.

Perhaps that's why prejudice grew rampant on the reservation. The Lakotas justifiably resented white presence on their traditional land. The white adults around me resented the various federal assistance programs. In a community where many of our white neighbors lived an existence just above the poverty level, farmers resented media portrayals of the destitute American Indians on reservations throughout South Dakota. The Indians around us always seemed to have enough money for new clothes, new vehicles, more cattle. Those who didn't, according to the voices of my childhood authorities, simply drank their money away and were, therefore, not to be pitied.

More than the money, though, my parents and grandparents seemed to resent the American Indian activists of the 1970s and 1980s who called for returning Indian lands to the tribes. By this time, the German-Russian settlers in our part of South Dakota had farmed their lands for two or even three generations. Now, suddenly, Indians and their big-city sympathizers were saying that these families should leave these lands and go find homes elsewhere. Some even called for the return of the buffalo to the prairies where we raised our cattle herds. Simply level the farms and fences built by settlers' hands, they said, empty the plains of those other than tribal peoples, and allow massive herds of buffalo to run free.

For a group of people whose forebears had been exiled from the lands in South Russia where their families had lived and farmed for over a century, this threat of a new exile ignited rage. Wasn't this the United States of America, the land of freedom? Weren't these the homesteads where they had tilled the soil, planted the windbreaks, and built the barns right alongside their fathers and grandfathers? Weren't these the homes where their mothers had birthed children and tended the dying? There was an irony here, as the cycle of conquest seemed to be coming full circle.

At the heart of the controversy stood one question: Who owns the land? In American Indian tradition, the concept of land ownership is foreign and one of the reasons why white occupation of the land occurred so quickly. To the German-Russian settlers, land ownership represented insurance that an exile such as the one from the South Russian colonies would never happen to them again. Loving, tending, and owning your land became the third great commandment, after loving God and neighbor.

But "neighbor" never extended to those living in the reservation towns nearby, I reflect now as I watch the riders race in the arena below us. For a few moments, the beauty of the costumes, the agility of the riders, and the majesty of the horses sweep over me like a romantic tide, and I long for a return of the prairies to the way they were a century ago. To see a Lakota on his horse, loping up a hillside, or a village of tipis nestled along a creek would thrill my soul. But one look at the well-dressed women beside me reminds me that that way of life would never return, no matter who owned the lands. For better or worse, American Indians in the twenty-first century are in most ways as far removed from their ancestral ways of life as I am from the South Russian colonies. It is one thing to celebrate customs with festivals and powwows; it is another to go back to a way of living marked by unrelenting hardship and personal danger.

In the years since I left South Dakota, changes have been happening that signal a healing between the Indian and white populations living there. Instead of celebrating Columbus Day, traditionally a holiday that marks the "discovery" of the "New World" by Europeans, South Dakota celebrates Native American Day. The idea of the "Buffalo Commons,"—returning the prairie to the buffalo—which so threatened white ranchers fifteen years ago, has evolved into more of a small-business movement than a forced exodus from the plains. Schools are more integrated; my sister, who teaches in the same high school we attended, instructs as many Indian children as white children now.

I don't exactly know the reasons for this nascent healing—there are certainly many prejudices still to be erased, and racial tensions in South Dakota may never completely die—but I suspect the healing comes at least in part from the recognition that both Indians and whites share an intense love for the prairies of South Dakota. Ironically, it may have been the very threat of losing their farms that caused the whites to empathize with the tribes.

There are many ways of owning land. One way is to pay a sum of money and sign a deed. I own no land; no piece of property belongs exclusively to me, or to my husband. But we own the places where we've lived in other ways: we own the small square of ground in the Boulder River valley where we buried our dog, Spud; the old homestead site where we made love under the cottonwoods while our horses grazed nearby; the forest cabin where our third child died in a rush of blood. We own the home we live in currently, even though we'll leave it when we leave this job. We own it because our childrens' laughter enlivens the air, because our sweat wrestles a garden crop from the clay soil each summer, because our mares drop their foals each spring in the pasture next to the driveway.

My family owns our farm in South Dakota for the same reasons—not because my grandfather paid for it (although he did), but because he, my uncles, my dad, and my brother have worked that land for nearly one hundred years. Because my grandmother, aunts, and mom have raised sixteen children there. Because my siblings and I bring our own children there, to learn to love the call of the pheasant and the palette of the sunset.

I understand why American Indians have had enmity for whites for so long; if someone took something so beautiful from my people, I would not be able to forgive them. The truth, which my ancestors did not want to admit, is that the tribes *did* own the prairies, in a way so profound that the white conquerors could not understand it.

Carmen has moved to sit on the other side of me now, next to the Indian women. The rodeo clown's antics between events prompt giggles from her, and the women laugh along. Her chocolate-brown eyes sparkle with the same delight that shines in their ebony eyes, and she does not realize that these women are of the same blood as the rider she was afraid of only moments ago. She sees only two women, who are laughing at the same things she finds funny. She accepts them for who they are, and they accept her.

And I am reminded of another child, long ago, who thought her sister was an Indian but didn't care. She was still her sister, after all.

Early April sun warms my shoulders as I walk to the corral and untie my horse. Shawn saddled her while I was dropping our three-year-old, Maria, off at the babysitter's. Now, he finishes cinching up the gelding he's riding today—part of the ranch string, not one of our own. Finally ready, we walk the horses through a tunnel that passes under the highway to the calving pasture. Today we are sorting pairs, pushing the bull calves and their mothers into one pasture, the heifer pairs into another.

"Need some help?" Shawn asks, kneeling beside Milly. The asking is only a formality. As usual, he kneels on one leg, offering the other as a sort of step-stool so I can reach the stirrup. I always feel a bit guilty, standing on him, so I step as lightly as I can, find the stirrup quickly with my left foot, and swing in. "Didn't know she was going to be such a tall horse when I gave her to you," he laughs.

"You should have thought of that—it might have saved your leg some," I banter back. Milly was my engagement present ten years ago, just a scruffy-looking weanling filly. Now, at eleven years, she stands tall and proud, her pregnant belly swollen beneath my own. "This is probably it for a while, Mil," I say as I pat her neck.

"You're right; she's getting close, isn't she?" Shawn remarks as he adjusts his horse's bridle. "And then after she nurses that foal all summer . . . "

"I'll be too close to ride anymore," I finish. Both Milly and I are expecting our fourth babies soon. I watch Shawn circle his gelding expertly before he swings into the saddle in one easy, fluid movement. Nerves flutter in my stomach, and Shawn hears my deep breath.

"You OK?" He looks over at me, reaches out to take my hand.

"Yeah—just a little nervous." My typical response. Despite being 56 raised around livestock, I am no cowgirl. In fact, this will be the first time I've helped to sort pairs. I've learned most of the little I do know about handling cattle by sitting back and letting Milly do the work. When we move cows and a wild one cuts away, I just try to stay mounted as she races after the cow and nudges her back to the herd.

"Just stay with me—you'll do fine," Shawn reassures. Having done this work since boyhood, he doesn't understand my nervousness, but he's learned not to disregard it. He put special effort into training Milly so that she would be calm and quiet, bolstering my own confidence.

After a few false starts, Shawn sorts off a pair and we begin moving them toward the bull-calf pasture. The little black calf, probably a day old, stumbles, confused, and the distraught cow bellows at him every few steps. "Oh, quit bellerin' at him and he'll come," Shawn says. This particular chore is slow, painstaking work, the kind that annoys Shawn the most. While he, like most cowboys, can easily move several hundred older pairs many miles in just hours, sorting just twenty-five to thirty newborns is a full day's work in this pasture. The terrain—with irrigation ditches, a swamp, groves of horse-high willows, and that highway tunnel—slows the cow and tires newborn legs quickly. Still, Shawn's impatience with this particular cow sparks empathy in me; we females, of any species, are so often expected to adapt to motherhood naturally, to have no misgivings and make no mistakes. I would probably bellow, too, if someone were trying to herd my baby and me.

We have trouble with that first pair. The calf keeps turning back, confused about where his mother is, and the cow cuts down the fence line when we finally get them to the gate. Eventually, however, we put them into their new pasture and go back for more as the calf suckles hungrily. From that point, the process flows more smoothly. Mother cows moo softly to their babies, teaching them to follow, and no more try to cut back at the gate. Milly and I take a few pairs out by ourselves. She tracks the cow, anticipating every turn, and gently nudges them toward the gate. I'm pretty much along for the ride.

Milly is my third horse, although I share ownership with Shawn on all but one of our twenty-odd head—the beginnings of a horse business we've been building for years. But only Milly, and the two horses of my childhood, Misty and Blaze, make up the small group of horses I've truly considered my own.

I shared my first horse, Misty, with all of my siblings, particularly my older brother, David. He woke me one morning when I was four, saying, "Do you want to learn to ride Misty today?" We haltered our little brown-and-white Shetland pony, and I spent the morning going around and around the pen behind the barn. Misty never wore a saddle, and I don't even remember a bridle—just her red halter and a lead rope looped around her neck. Thus, she didn't always rein well, but we loved her. She'd stand stock-still and let my sisters and me slide off her rump into a pile of hay.

One day, though, she didn't stand still. It was a beautiful summer day, so David and I decided to ride her out in the pasture. As he was off shutting the gate, Misty bolted, with me on her back. Her pony legs couldn't carry her very quickly, but she did outrun David. Somehow I ended up on the ground, though I don't remember how. Maybe she turned abruptly, or maybe I just jumped off—a dangerous habit that would get me into trouble years later. Relieved of her burden, Misty trotted off to find the other horses. I wasn't at all hurt, so David and I decided to try to catch her. After a few hundred yards with no luck, David said, "Hey! You want to go down to Dwight's to get a drink of water before we go home?"

Dwight and Linda neighbored our place to the south; their house stood about a mile and a half from ours. We could easily have turned around and walked home for the water, but instead, we trekked south through the pasture, past the old corrals and the windmill, to the railroad tracks separating the two farms. Just as we neared the tracks, a train hurtled past, and we crouched behind a large sandstone boulder, listening to the thunder of the wheels. After it passed, David told me something about the boulder: that it wore away a little with each season, until someday there would be nothing there. At the time, my childish understanding thought things so solid never changed. The rock would always be there; David and I would always share our love of horses; and I would always be the "little brother" David never had. Surely age and gender would never separate us, just as surely as wind and water could never change that rock.

We reached the neighbors' shortly afterward, took a drink from the garden hose (they weren't home), and headed back. As we neared our place, we saw Dad's truck parked on a hill overlooking the pasture. Evidently, he had come in from the field to search for us. Our parents scolded both us, and probably punished David. That was our last summer adventure with Misty. By the next year, David was old enough to work in the fields with Dad. The following winter, Misty died of electrocution when a power line fell into her water tank during a storm.

Milly and I head back to the main pasture now and meet Shawn coming in with another pair. He's riding a stocky dark-brown horse that the ranch bought from my brother last fall. I'm always a little surprised to see him on any horse other than his sorrel gelding, Chance. Shawn usually rides Chance whenever I ride Milly, but he makes time to work with all the ranch horses, and today it's Snort's turn.

I don't consider Chance one of my horses because Shawn owned him before we ever met. When Chance was born, part of a band of horses running wild in the Sweetgrass Hills of northern Montana, his mother still nursed a yearling foal. Consequently, the newborn Chance nearly starved. Shawn's boss, a horse trainer, bought the entire herd and found the baby almost dead. He gave it to Shawn, saying, "If you can save him, you can have him." Shawn took the runt home and bottle-fed him. Thirteen years later, he still rides, ropes, and relies on Chance.

"How's it goin'?" Shawn calls out as we approach.

"Fine; Milly's working great."

"Well, you sure look good up there. Want to take this pair in, and I'll go get another?"

"Sure." Working separately like this doesn't allow much time for conversation, but as I ride off, I'm pleased by his compliment. Though I'll never feel like a true cowgirl, I like hearing Shawn say I'm riding well. Like most times I ride, though, I find myself wishing David could see me, to see how far I've come.

David wasn't living at home the day I had my accident, the day my confidence shattered. It happened while I was riding Blaze, a four-year-old filly my dad had purchased for my sisters and me when I was thirteen.

David trained Blaze at the corrals he leased, since he had moved thirty miles away from our farm after he married. When he brought the filly back, he re-taught me many of the lessons I'd forgotten since Misty died: how to saddle, how to mount, how to keep control of the horse. Then, he admonished, "She's just green broke now; she needs a lot of riding. It's up to you to keep her rode down, because nobody else is gonna do it." For several weeks that summer, I rode patterns in the arena behind our barn. I especially practiced the cloverleaf pattern, thinking I might try barrel racing the next year. Then, after I felt more comfortable with Blaze, I began riding her outside the arena.

By fall, we had developed a routine: I'd quickly change my clothes after school, saddle her, and ride her out into some unused pasture. We trotted at first, but as soon as we hit the old cattle trail, I'd crouch down low over her neck, squeeze with my thighs, and let her run. We flew over the hardened trail, her hooves drumming a rhythm that sounded the passion welling up in both of us. Together, a four-year-old filly and a fourteen-year-old girl, we raced toward the changes ahead, changes that would eventually separate us.

I remember that Blaze opened up the masculine world of cowboying for me. My dad, raised in a strict, German-Russian home where women did not do "men's work," relied solely on David to help in the fields. But that fall, with his son gone and the work remaining, Dad allowed me to help him move our small herd of cattle. He rode point near the front, dignified on his copper-bay gelding Duke. Blaze and I brought up the drag, the traditional position for inexperienced riders. Still, I cannot describe the pride I felt to be working with my dad.

The next spring, however, my cowboying career ended. We had a long winter, and I didn't ride Blaze for several months. David brought her in one Sunday in May and told me to exercise her in the soft stubble fields east of the house. The first time I rode her, I quickly discovered that her strength had surpassed mine over the winter. When I nudged her into our run, she jerked, pulling rein from me until I had too much slack. I panicked and pulled back on what rein I had left in my hand, but to no avail. She just kept running, taking more of the rein with each lunge of her head. Then, more terrified of being on a runaway horse than on the ground, I jumped to the soft earth. She reached the corrals, and Dad was just coming to look for me when I walked into the yard. "What happened?" he asked.

I couldn't tell him that I had jumped off, suddenly too scared to ride out her run. Ashamed, I mumbled something about falling when she turned too quickly and went to the barn to unsaddle my horse.

The next time, on a sunny morning a few days later, I didn't let Blaze run, but she acted nervous and twitchy, fighting the bit and tossing her head. As we turned toward the barn, I sensed my control slipping away. My youngest sister was riding her bike in the driveway; I called softly for her to get out of the way just before Blaze bolted. Snippets of Hail Marys floated through my head as the pasture fence loomed closer. Fearing she would try to jump it, I again flung my body out of the saddle. This time, though, my boot caught in the stirrup, and the ground ripped at my hands, arms, and face as I struggled to kick free.

Dad reached me first, asking me if I could wiggle my toes, then helping me stand. Blaze stood panting at the fence where she had stopped, tail held high for the geldings watering there. It would be years before I understood how her changing passions had made her run for more than speed that week.

At the clinic, David came to see me. My injuries—pulled muscles and scraped skin—amounted to little compared to the broken bones he'd suffered on horses and bulls. Still, because of my gender, perhaps, or my inexperience, my parents decided I could no longer ride Blaze unless either David or Dad were present. David reassured me he would find time to work with me, but his adult commitments pulled him farther away.

That afternoon, I limped to the corral to water Blaze. She had calmed, and she nuzzled my back as I led her through grasshopper-infested grass to the tank. As she drank, I laid my head against her neck. "I'm sorry, Blaze," I whispered. I knew, even that day, that I wouldn't ride her again, that I would always be too afraid of another runaway. Blaze was only four, too young and green to know better than what she had done, and I had failed her, failed David, failed myself. Dad turned her out to pasture the next week, and sold her the next year.

By now, Milly and I have our last cow-calf pair of the afternoon; soon I need to meet my older daughters' school bus. Suddenly, the cow breaks right, all but forgetting her calf. Milly charges after her. Panic rises in my throat; will I always be fourteen, riding across our farmyard, when I feel my horse bolt under me?

We catch up to the cow as she turns back to look for her calf, and soon we're trotting gently toward the gate again. I calm down, but chastise myself: to think that the speed that once thrilled me now terrorizes me, even on Milly.

Six years after my accident on Blaze, I met Shawn. Bent on impressing him, I insinuated that I could ride most any horse—never mind that I had rarely sat in a saddle in six years. We fell in love, and just before he left at the end of the summer, we trailered two horses to the mountains for an overnight pack trip. As we wound our way to the alpine lake we wanted to see, I asked the age of Tomcat, the gelding I rode. "Oh, he's a two-year-old; I've been breaking him for a friend, but I haven't had much time to work with him this summer," he replied, as casually as he might point out a wildflower or an elk rub.

The truth about my riding ability became apparent when not only Tomcat, but also Tomcat's rider, panicked at every stream crossing and spooked at every chattering squirrel. Shawn lost patience once, when I called to him for help as the horse balked at stepping over some deadfall in the trail. "Just make him! I can't lead you—I'm on a colt, too!" he barked, and my tears came quickly. At camp later, I told Shawn the whole truth, all the while fearing I would lose this cowboy I loved. After all, didn't he belong with a cowgirl whose skills matched his? Since the accident on Blaze, my brother and I had drifted apart; would I also lose Shawn because I couldn't ride?

Thankfully, Shawn saw that behind my fear the same passion for horses still burned. He promised to help. From then on, he offered me only the most gentle, trustworthy horses he knew—no more Tomcats! Still, when he presented me with Milly the Christmas after we were engaged, I almost refused to accept her. She was too beautiful, meant for a real cowgirl. Besides, what if I failed her, the way I had Blaze? But Shawn believed—as David had years before—that I belonged on good, powerful horses, not ponies or nags. I saw that I had to start moving past the fear.

One September afternoon before our engagement, on a weekend trip to Idaho with two girlfriends, I walked along a gravel country road until I spotted some horses grazing nearby. To my delight, I saw that one of them was a Shetland pony, marked similarly to Misty's tobiano pattern of brown and white spots. I scrambled up the embankment to the fence, where a crab-apple tree held the last of the year's fruit on its more hidden branches. Breaking off a few, I offered them to the horses, particularly to the pony. He accepted it nonchalantly, as though strange young women fed him crab apples all the time, and slobbered over the fruit as I ran my fingers through the impossible tangles in his mane. By any horseman's standards, this pony was ugly: sway-backed, hooves overgrown. Still, he thrilled me, awakened in me the little girl who loved nothing more than to ride her shaggy Shetland pony. Ponies don't buck, and if they do, the rider doesn't have far to fall.

On the return trip to Montana and to Shawn, my friend's car climbed glorious mountain passes that awed us with autumn beauty. I kept thinking about that pony, about wanting to feel safe and secure the way I had on Misty all those years ago. The changes rushing at me that fall—my dad's worsening health, Shawn's impending marriage proposal, my need to decide a career path—both frightened and excited me. I wanted to slow down, to jump from the runaway, and yet, at times, the speed was exhilarating. Was this what Blaze felt when her overwhelming passions pushed her out of our familiar routine into desires she didn't understand?

At the top of one mountain pass, I thought of my uncle Fritz, whom I never knew. Although all of my nearest ancestors have been farmers, there runs through the Lipp blood on my dad's side a strong horseman tradition, the pull of horses that some, like David, have not been able to resist. Fritz was a cowboy, too, and he died on a mountain pass like this one, struck by lightning when a sudden Colorado thunderstorm blew up as he moved cattle. He must have known what might happen to him during that storm, and yet he didn't dismount but stayed with his horse. The more settled of my relatives—mostly the women—used Fritz's death as an example, a warning not to be outside with horses during a lightning storm. But my dad rarely mentioned the death of his closest brother; I think he recognized that dying horseback in a spectacular thunderstorm atop a mountain is just about the best way for a cowboy to go. It certainly beat slowly losing your independence as an insidious disease steals your lung capacity.

Suddenly, I realized I could never ride to the top of this mountain on a pony, safe and familiar as it might be. I had to move forward into my life, with all the beauty and tragedy it held. And I had to learn to ride again.

The learning to ride again, and to trust my ability, is simply a journey. I've suffered setbacks, paralyzing moments when I've again longed to jump to the safety of the earth. I realize that my womanhood is part of the reason for my terror. Falls simply don't mean the same thing for men, like Shawn and David. Men risk bruises and broken bones, and occasional accidents like Fritz's bring death swiftly. I risk not only these same things, but the death of the child within me, as well. My girlhood limited me after my accidents on Blaze; now motherhood has altered my future as a horsewoman. Because I am a mother, I require equipment most men don't think about as they saddle up: a babysitter to watch over my kids, and a well-trusted horse to watch over me.

In this respect, my admiration for Milly grows. How does she overcome her natural tendency to go slowly, to rock the foal within her gently? How does she override her fear of falling on this muddy, slick ground in order to burst after a runaway cow? For she truly is as adept as any of the geldings Shawn rides; being female is not a handicap for this glorious red horse. Motherhood for her simply brings the expected limitation: long summer months each year when I can't ride her, when I choose not to, because she nurses a foal.

But when we have been able to ride, Milly has carried me through glorious moments. On Milly, I've ridden through ochre and pink badlands; I've climbed up craggy boulders to a nearly inaccessible mountain lake; I've trotted down off a mountaintop in pouring rain; and I've been welcomed into the bond of motherhood in a quiet barn at midnight, the smell of birth and hay mingling with her foal's high-pitched nicker. Milly, and Chance, have learned to travel side by side so Shawn and I can hold hands while riding, and they have grown accustomed to grazing quietly nearby while he and I have lain under the pine and cottonwood trees. She's been the loaner horse for the rare times David has visited us and the nanny horse for my own daughters as they've learned to ride. As we return to the corrals and dismount, I nuzzle her neck, hugging her the same way I once hugged Misty and Blaze. What a gift this incredible creature has been to me.

It's 2007, six years after that April pair-sorting ride. Today we help David brand his calves, fighting first the mid-June South Dakota heat and, later, an imminent lightning storm. Shawn ropes the calves on a succession of young colts, none as reliable as Chance or Milly, until the wind finally takes his loop away from him. Panels are moved to create a smaller pen, and the men begin wrestling the last twenty or so calves to the ground. David asks me to ferry the hot irons from the stove to him, since he is now working inside the pen. It's a decision of necessity—he has few other helpers—but still, he chooses me. In his eyes, so much like mine, I read the trust: he knows I will watch which calves go down, grab the right irons, and get them to him quickly.

Lightning crackles in the air, sometimes blinding us. We all, having been conditioned, think of Fritz, and every now and then someone mentions the danger. But I don't think of quitting. The storm isn't the only electricity in the air tonight. I'm helping to brand my brother's calves, continuing a legacy of helping each other that began that summer morning on Misty. The wind and water of our ages and genders *has* worn away our childhood bond, to be sure. But equally true is that we are tied by other, less quantifiable, energies: the loss of our dad, our shared fascination with horses, and our deep love for one another's spouses and children. The sandstone boulder may be gone, but the memory of a brother and sister crouching behind it remains, indelible.

I rode Milly just last weekend, on a short mother-daughter ride with two friends and their girls. Maria, now nine, was mounted behind the cantle on Milly's rump. She wanted to lean forward, arms clutched around my waist, but I made her sit up tall, use her strength to find balance. "Mil would never do anything to hurt you," I reassured. In old-time cowboy lingo, Milly would be referred to as a "broke" horse, meaning broken of wildness, of any behavior that could threaten the rider. And I sense that Milly is "broke," although I prefer to think of her as well-trained. Her teaching has not so much defeated her as opened her up to her potential as a horse.

However, Milly is "broken" in other ways. For one, animals don't get to keep their babies and watch them grow, the way women do. Over the years, I have watched enough mares run the fence, whinnying for their babies, and listened to enough mother cows, bawling in the weaning pen, to disbelieve any notion that animals don't know or care when they are separated from their young. That pain of separation is another bond my horse and I share; eleven years ago, I lost a child, too, and that brokenness never leaves me. Maria, however, would not have been riding behind me if that child had not died.

By coincidence, both Milly and I had sons that year of the sorting. Mine, now almost six, waited back at the house for me to complete this ride. Hers, our stud horse, had been killed by lightning just days earlier. Thus, neither she nor I were in the mood for the camaraderie of a group ride, and we turned for home early. "Do you think she knows about Little Doc, Mom?" Maria asked.

I thought of all the things Milly knows: which rider is on her back; how to watch a cow, pull a rope, cut a calf; what paths to take up and down rocky slopes; how to nurse and protect a foal; and, most of all, how to heal and give courage to a woman who once had none. Did she know that her son had died? Perhaps not, but perhaps yes. "I think she does, Honey," I responded. "She's Milly." And my beautiful red-dun mare trotted on, carefully tending Maria and me, carrying us home.

The tomatoes are ripening late this summer. Perhaps the drought delayed them, or maybe I planted too late. I know my Grandma Quinn would have chided my end-of-May planting; her tomatoes would have been in the ground as soon as it warmed, usually by Mother's Day. On this mid-August morning, I wander through the bed, lifting vines to search for hints of red, and I worry that the still-green fruits will all ripen in September, when I'll be too pregnant to harvest and process them. As I search, I notice many cucumbers that are ready for pickling, though, so I plan to take care of those this morning.

Laura, who's tall and strong for her eight years, picks the cucumbers and carries the large bowl to the sink. Being seven months pregnant makes crawling on the ground difficult for me, so I appreciate Laura's help. I appreciate all three of my daughters; the girls' extra caring and compassion is welcome this hot summer. This has been a tough pregnancy, and I tire easily.

Making pickles, however, isn't tiring. The job is much simpler than canning those tomatoes will be. I slice the long, thin cukes while a brine of sugar, vinegar, and water boils on the stove. Into each sterilized jar I pack a slice of onion, a toe of garlic, some dill, and enough cucumber slices to reach the rim. Then I pour in the hot brine and fasten the caps. The jars go into a pot of boiling water for ten minutes—and that's that. In less than two hours the job is done.

Later in the day, I rise from a nap to feel a trickle of fluid between my legs. It stops, though, so I ignore the feeling. Throughout the afternoon, however, I keep sensing more trickles, usually whenever I stand up or bend over. *Could I just be wetting my pants?* I wonder, disgusted with myself. I visit the bathroom to empty my bladder repeatedly, but the fluid continues. When Shawn comes in, we de-

cide that, although it probably isn't amniotic fluid, I should go to
the clinic just to be sure. He goes back to work and I drive to town
with the girls buckled in the back seat.

Within an hour, I am hooked up to a fetal monitor, frantically
trying to reach my neighbor, who will babysit my girls and send her
husband to search the hay meadows for Shawn. The doctor tells me
I "broke my water"—as though I did this on purpose—and must be
flown to a large hospital in Billings, Montana, where specialists can
care for the baby and me. I am not even allowed to drive my daughters back home. They stand, crying and frightened, by my bedside.

When Shawn arrives, dirty and sweaty from being in the field,
the look on his face is pure terror. We've already lost one baby, a
miscarriage five years ago; will we lose this one, too? The nurses
ask if I have a history of premature birth, of smoking, of any behav-
iors that would put the baby at risk for early delivery. No, no, no, I
tell them. I've always been healthy; my other full-term pregnancies
were healthy; there has been no indication that this could happen.

And yet, it *is* happening. My baby is about to be born seven
weeks prematurely, and its survival is not guaranteed.

I don't know much about the circumstances leading up to the
premature birth of my Grandma Quinn's little girl. My mother has
told me Grandma slipped on the ice that day, helping to milk cows.
Whatever the reason, she wasn't flown to any major hospital or
placed under highly specialized care. She and Grandpa, alone and
afraid, faced delivering their baby at home. I imagine my devout
grandmother praying the Rosary between contractions.

I don't have the presence of mind to complete a prayer—maybe
Grandma didn't, either—so I call my mother. She and I share the
bond of having lost children, and she hears the fear in my voice. I
ask her to pray. A part of me believes my mother's prayers count for
more than my own; she is so much more faithful, so much a better
person than I am.

Shawn and I arrive in Billings; labor has not started yet, a good
thing. Nurses promptly start an IV line to deliver the drugs to stop
the slight contractions I've been having. Our neonatologist comes
in to do a sonogram. Up until now, we have not known the sex of
this child, but the urgency of this situation makes us want to know.
The doctor points out the scrotum and penis, and Shawn begins to

cry. He's had this child's name—Cody—picked out since before we were married. Now, Cody may not live past his delivery.

As during other periods of grief in my life, I am overcome with the wish that I could simply turn back time, back twelve hours or so, to "then." *Then*, everything was normal. *Then*, our son was not threatened with leaving the safety of the womb before he is ready. *Then*, my absurd worries centered on whether enough guests would come to my four-year-old daughter's birthday party this weekend and whether the tomatoes would ripen in time for me to can them. Now, I would gladly trade places with my son, let my own life be threatened so that he would have a chance, and so that I won't have to see that look on Shawn's face.

But I'm no saint: any mother would trade her own life for her child's. I am certain my grandma would have. However, Grandma didn't have these drugs to halt her labor, and her daughter, named Lavonne, arrived three months prematurely. The baby fit into the palm of my grandfather's hand. During her life, Grandma would often talk about how small, yet "perfectly formed" the baby was. Grandma remained quietly, prayerfully pro-life until her death; the image of Lavonne never left her.

On Friday morning, twelve hours after our arrival, a nurse guides my husband, my best friend, Kelley, and me on a tour of the neonatal intensive care unit. From my wheelchair, I peer up at isolettes holding babies tinier than any I have seen. They are alive, though monitors register their every breath and heartbeat. I read one baby's story on a notecard taped to the end of her bed: born two weeks ago, at twenty-four weeks' gestation, the little girl weighed only one pound, six ounces.

I imagine that this little girl weighs about what my aunt would have. Yet Lavonne breathed only a few minutes before dying; this baby has lived two weeks. Our doctor tells us that Cody probably weighs around three and a half pounds. Two steroid shots administered the night I arrived have helped his lungs to mature. Already our baby has far better chances than my aunt did.

After Shawn returns home to pick up the girls, I finally allow myself to pray. I know the anger I will feel if Cody dies. Shawn has waited so long for a son, and the girls have been so aware of this pregnancy, ever since we shared the news with them six months ago. To take that away from all of them . . . God wouldn't do that, would He? I haven't always been the best mother to the children I do have; maybe *I* don't deserve this little boy. Still, Shawn and the girls don't deserve that kind of grief. But I know life isn't about fairness—otherwise, women like my grandma and others much holier than I would never lose children.

I came to know Grandma Quinn—Ida Louise Brockamp Quinn—relatively late in her life, when she was in her sixties. I don't remember ever not knowing her; her home, three miles from ours, was a permanent fixture in our weekly routine. At the very least, we kids visited twice a week—always on Sundays, and some other time just to say hello. The plump, jolly Grandma of my early childhood grew thinner and more frail as she aged and battled cancer, heart disease, and stroke. Yet despite her struggles—and this trait is common to so many of my ancestors, to the people of that era—her cheerfulness never diminished. She busied herself in summer tending immense vegetable and flower gardens, and in winter knitting sweaters and afghans. She often remarked how she enjoyed these pursuits because she could work and pray at the same time.

My mother calls the hospital once every day, sometimes twice. The stories of her older sister's death, plus other near-tragedies with premature babies over the years, have frightened her. Mom knows what it is to lose a child. She repeats during each phone call how she is praying for the baby. Once again I am amazed at the faith of the women in my family. Through miscarriage and stillbirth, Mom lost five children. One for each of us still living. And yet she turns to God in complete trust.

The wounds from my own miscarriage five years ago heal slowly. Earlier this summer, we drove past the camp where I lost the baby; by pure coincidence, our travels took us to that part of Montana exactly on the fifth anniversary of the death. Shawn and I held hands, and I caressed my once-again round tummy. One morning last March, when I was newly pregnant with this baby, I woke to find spots of blood. Horrified, we thought another child was dying. I questioned God, begged, raged. Two weeks after the spotting mysteriously stopped, the doctor found our baby's heartbeat. Overjoyed, I began carefully plotting my meals for maximum nutrition, slowed down my exercise to walking and yoga, and took lengthy afternoon naps. Despite numerous reassurances from family, friends, and physicians, I'd never stopped wondering if neglecting my own health had caused our third child's miscarriage.

Now, death—or at least disability—threatens our son. All through this pregnancy, an inexplicable fear has nagged me: what if something goes wrong? What if I haven't been careful enough? And here we are now. I fall easily into guilt: perhaps I shouldn't have made those pickles, perhaps I should have rested more, perhaps . . . but my logical side knows better. Although the old wives' tales about work causing premature labor are thrown at me often, I know nothing I did caused my water to break. Sometimes, no matter how carefully the tomatoes are planted and nurtured, they still ripen slowly, or not at all. I am learning what Grandma Quinn knew: gardens and babies, and most other good things in our lives, are completely out of our control.

Grandma's faith required her to trust God's plan, no matter how much pain life brought. When she lost her first child, I doubt whether she could see the reason behind the despair. But she trusted, anyway, all through the subsequent births of eight healthy children. Perhaps God allowed her to experience the agony of losing Lavonne so that, some thirty years later, she could comfort another daughter, my mother, through her own four miscarriages and the stillbirth of my older sister.

The contractions intensify during the next week, and my doctor puts me on a regular schedule of medication to stop labor. I worry about infection and the little remaining amniotic fluid but am told often that Cody remains healthy. Still, I listen intently to the baby monitor, and my stomach drops each time his heartbeat slows. My doctor, a quiet, calm man with a gentle smile, explains that he hopes I can carry the pregnancy to week thirty-five—two weeks away—but that "everything is in God's hands now." I am grateful to be in this hospital where nurses stop by to chat, the doctor's assistant takes me for wheelchair rides, and the doctor himself shows up with a cup of candy when four-year-old Maria has to celebrate her birthday in my hospital room.

Tomatoes require various conditions to grow and, especially, to flourish: rich soil; deep, regular watering; hours of sunlight;

and protection from wind, weeds, and pests. I once thought babies depended only on the mother's proper care and nutrition to be born healthy. Perhaps that's true for most babies. But certain special ones—like that little girl in the isolette, like Lavonne and Cody—must have help from extraordinary professionals who are assistants to God.

Today is August 30, 2001, the day before our tenth wedding anniversary. I spent most of yesterday in a labor room because irregular contractions came all day. They stopped by evening, but today a different pain grips my lower back every ten minutes, and so I am wheeled back into a labor room. Shawn drove up yesterday, after helping Laura and Carmen off to their first day of school. We all wept when they left the hospital after their weekend here. They knew I wouldn't be there to lay out their backpacks and braid their long hair, to take first-day-of-school pictures and smile enthusiastically as Carmen began first grade, Laura second. Instead, I talked to them on the phone amid more tears. At least my youngest sister, Michele, has come to stay a few days with them; I pray that our shared upbringing will prompt Michele to mother them the way I would.

The pains in my back intensify, now clamping down on my abdomen as well. I know this is the day, but the doctor, my nurse, and Shawn don't think so. After a long morning, an ultrasound at noon shows that my cervix is still closed. I am both relieved and dejected—relieved that our son remains safely inside my womb, but dejected that all this pain has been for nothing. Certainly I should be willing to endure the cramping so that my son avoids danger, but I am just so tired. As always, labor makes me feel weak and weepy. I tell Shawn, "I can't do this. I'm not strong enough." And, as always, he answers, "Yes, you can. You have to."

What terror must have gone through Grandma's mind during her first labor. Not only did she experience the mind-blowing pain—a pain no woman is ever prepared for—but she must also have realized that the labor would end tragically. In those days, most premature babies simply did not live. To go through all that pain only to lose the baby. . .

My own miscarriage felt just like labor, only shorter. So did my mom's. And she went through a normal labor when my sister was born, strangled by her own umbilical cord. Anger is our natural response to injustice. How can a woman not be angry at God after such supreme injustice, after being cheated so?

In the afternoon, my body feels feverish. Shawn has changed his mind and now believes it's time, too. Around three o'clock, my nurse comes in to find me crying. Surprised, she asks what's wrong. "It just hurts so much," I say. She leaves the room, comes back shortly with a pill. "The doctor says you can have this, but that I have to check you first." She reaches inside me to feel my cervix, her eyes widening as she exclaims, "You're at three centimeters!" The rest of the afternoon blurs into intervals of pain and release. I seesaw between anticipation and dread. What if something goes wrong?

Labor progresses quickly, and I feel the pressure of Cody's head almost before I am ready for it. People are crowded all around my bed as I push my baby from safety to uncertainty. I reach down and pull from my body our son, who squalls loudly and pees on my leg. The gratitude and joy I feel go beyond words.

We hold our son briefly—his tiny body is not even as long as my forearm. He seems all head; his head is normal-sized for a newborn, making the rest of him seem tiny by comparison. Soon a nurse takes him away to the neonatal intensive care unit (NICU), his home for the time being. The neonatologist returns shortly to tell us that our son weighs four pounds, six ounces—a good weight for his age—and is healthy. "He probably won't have to stay long," the doctor smiles.

Shawn leaves me then, to go to Cody. I'm finally able to join them after the anaesthesia wears off. Although we toured the NICU last week, actually seeing our son there shocks me. He lies on an elevated bed under heat lamps, three wires taped to his tiny chest, an IV tube inserted into his two-inch-long foot. When the doctor said Cody was healthy, I thought he wouldn't need all these wires and monitors. "Healthy" becomes a relative term in the NICU.

A nurse explains how each monitor measures pulse, respiration, and oxygen level. She advises us not to watch them too carefully— "watch your baby instead"—but it's hard not to pay attention to the beeps and electronic lines. Eight years earlier, I had been watching the heart monitor when my dad died and saw the pulse line go flat even before the alarm sounded. Now, I peer at each reading, trying to anticipate any dangerous blips that might signal my baby's death. He sleeps naked except for a diaper, alone on that bed. He should be swaddled in flannel, nestled in my arms. He should still be inside me. The nurse, a gentle, tall woman named Diane, asks if I would like to hold him. I sit in a rocker, take him from her, and instantly feel a wave of emotion overwhelm me. She brings a privacy screen so I can sit in the chair and cry.

When I wake the next morning, my hands travel automatically to my abdomen, which feels strangely flat. Panicking, I look for my baby, then remember that he sleeps down the hall. How horrifying it must be to feel that strange flatness and know your baby has died.

We expect a natural order in life. When we plant the tomato seedling in spring, we expect to harvest red, ripe tomatoes in late summer. When we conceive a baby, we expect to give birth to a healthy infant nine months later. When life doesn't proceed according to our expectations, we are shaken. But true character grows from learning to cope with the thing we didn't expect, the thing we hadn't planned for. True character grows from living through the droughts, through the deaths, through the disappointments and shattered dreams. True character grows from losing a Lavonne and then having the courage to conceive eight more children.

I'm discharged thirty-six hours after Cody is born. I move across the street, to a home for parents of seriously ill children. After helping me shop for groceries and get settled, Shawn leaves again, this time to get the girls so they can meet their brother. The drive from Billings to our home is 170 miles, or about two and a half hours by interstate highway. Shawn has made the trip three times already.

Being away from the hospital after such a long stay disorients me. I wander through the house kitchen and television room, feeling lonely and small amidst so much space. Outside, the late-summer heat assails me, and I tire walking the half block back to the hospital.

I don't know what to do with my time, so I scrub up and sit in a rocker next to Cody's bed. He rarely awakens, even when, every three hours, I do his "cares": taking his temperature and changing his diaper. He's been moved from the open bed with heat lamps to a plexiglass isolette, enclosed so that heat doesn't escape through the top. I learn to care for him through two round windows in the side.

Cody doesn't know how to nurse; the doctor tells me that many "preemies" are born without a "suck-swallow-breathe" reflex and are hospitalized until they master feeding. I learn that Cody will remain here until he can regulate his body temperature, take all feedings by breast, and gain weight. While I'm grateful he has no other hoops to jump through, I'm dejected at the realization that simply learning to eat could take him several weeks.

In the meantime, he receives regular feedings of my pumped breast milk through a nasogastric tube that has been inserted through his nostril and down into his stomach. A syringe of my milk is attached to the end of the tube and depressed slowly every three hours. My days soon evolve into a routine of eating, sleeping, and caring for Cody—not so different from what they would be if we were at home. I want to resume some normal activities, so Shawn brings my writing notebooks and piano music from home. He also brings ripe tomatoes.

Grandma Quinn loved to feed people. Her Thanksgiving dinners overflowed with food. Her cookie jar held batch after batch of soft, molasses "monkey-face" cookies or moist, pillowy sugar cookies made with sour cream from her cows. Even as she grew weaker, she would bake a cake for Grandpa every week—usually a white velvet with fudge icing, a spicy nut cake, or a deep-chocolate with brown-sugar frosting.

For many, a love of gardening arises from a love of God and others. In Grandma's case, her love of vegetable gardening developed from a need and a desire to feed her family good, nutritious food. I reflect on this as I prepare a salad from the vegetables Shawn has brought. He's here with the girls again, and it feels good to prepare a meal for my family, even if our circle currently feels incomplete.

Cody grows stronger daily, although he still falls asleep so easily that nursing him is almost impossible. I learn to pat his back, flick the soles of his feet, talk, and sing while nursing—anything to keep him awake while he eats. We learn, too, of the "NICU dance"—two steps forward, one step back. Each time he has a good feeding, he'll be too tired at the next to nurse well. I suggest letting him determine his own schedule, but he is kept on the three-hour routine nonetheless. The nurses explain that he needs feedings this frequently in order to grow. The problem is that if he doesn't nurse well, he still receives milk through the damn tube; he is not learning to associate hunger with nursing.

The weekends are a bustle of activity, with Shawn, the girls, and other family members visiting. During the week, as they all return to school and work, loneliness and despair creep around the edges of my soul. I start each day optimistically, certain that Cody will make great gains. I end each day crying into the telephone.

In a garden, success comes in small steps: first the tomato seedling puts out new leaves, then more vines, then yellow blossoms that swell into green fruits. Success with Cody comes the same way: first he is more wakeful, then he nurses vigorously for two, even three feedings in a row. By the second week, I'm able to give Shawn good news each evening: a weight gain, fewer tube feedings, even the monumental move from an isolette into a regular crib.

Around us, the other babies in NICU have their own successes. I marvel at the patience of one mom who's been coming here every day for four months. Her son is four months older than Cody but weighs less. Another young mom spent two months with her baby boy in a Denver hospital before his transfer to Billings three weeks ago.

Despite the frustrations we all feel at being away from our homes and families, a peace prevails at cribside. We enter, pull a rocking chair to our babies' beds, and sit. I reflect that, strangely, perhaps this time in NICU is a gift. At home, with three other children, I would never sit and hold Cody for hours between feedings. Here I have nothing else to do. I wonder how many times Grandma must have wished that she had been required to put aside her chores and simply hold her little girl.

The eleventh of September arrives. I have spent the night "rooming in" with Cody, sleeping with him in a hospital room so I can do round-the-clock feedings. In an hour or so, I will speak to the neonatologist to see if Cody can be released today. I am at the sink, washing up to go to Mass in the chapel downstairs, when the television screen shows a plane flying into the second tower of the World Trade Center, the first tower already ablaze.

The doctor seems reluctant to dismiss Cody because his weight gains are still somewhat erratic. A nurse and I pressure him, saying that Cody will be under close supervision at home and will do fine. Today's horrible events have made being separated a stress my family can simply no longer bear. Cody and I need to be at home; Shawn and the girls need us there.

Through a long day of waiting for doctors, packing up, and watching terrible news reports, one thing centers me: Cody needs to eat, and I am his source of food and comfort. The cribside peace becomes even more precious. When the hospital intercom announces a disaster alert, I go immediately to his bed and stay there. If the unthinkable happens, our son will not die alone.

Grandma's daughter did not die alone; she died surrounded by the love of her parents. Grandma's peace arose, perhaps, from knowing that she had accepted tragedy and made the best of it. We don't have control of life; we have control only of our reaction to its events. That total surrender, that giving up our self and giving in to life is what transforms tragedy into sanctity. Losing her daughter broke Grandma's heart and drove her to her knees; when she struggled to resume her life after her loss, she was softer, stronger, saintlier.

We finally bring Cody home on the evening of September 11, 2001. Later that week, with the girls off to school and Cody asleep in his cradle, I can the tomatoes. By now I've lost much of the crop to birds, bugs, and early frost, but I'm still able to prepare a dozen jars each of salsa and spaghetti sauce. I look at my sleeping son, then outside at the bright September day. My harvest this fall was not what I had planned; it has become, instead, so much more.

Today marks the second Friday in July, and at 1:30 P.M., the thermometer on my deck reads 102 degrees Fahrenheit. I pour water on the pots of wilted flowers outside the door and then walk out to the garden to reposition the sprinkler, though I won't actually water until tonight, when the air cools some. The garden crops stand short and stunted: the water they receive from the sprinklers is not enough, too laden with minerals leached in from the ground. Nothing substitutes for rain, and we've had none since spring. I'm afraid even to pray for moisture anymore; hope unfulfilled scorches like the afternoon sun.

Compared to the plains of South Dakota, our valley is blessed: we have the power of irrigation to water crops, lawns, and gardens. Two creeks carry snowmelt down from the Big Horn Mountains to drain into several creeks in the Powder River Basin. Along the way, dams and ditches divert the flow to fields and reservoirs. A person actually has to travel to the brown hills where the cattle graze, has to hear the brittle grasses crunch underfoot and smell the dusty, parched earth, to understand the severity of this drought. The land has put on its late-August dress, tan and umber and beige, although we are not yet halfway through July.

Summer stretches interminably, the days filled with mundane tasks and demands from my four children. I long to stretch out on the couch in the dark basement, undisturbed. As I stave off the drought by dutifully watering my plants, I struggle to fight off depression by plodding through my list of daily chores. That list, however, like the sweep of brown grass advancing down the slopes, grows incrementally each day. Knowing all I "should" do as wife and mother oppresses me more than the heat.

This morning, giving in to my kids' pleas for some sort of outing, I took them on a bike ride to the Brug home across the highway. The place stands vacant most of the year, except when it is used as a hunting lodge in the fall. Originally, however, a German-Russian family named Brug occupied the tiny frame house beneath the cottonwood trees. I remember my surprise last year when I discovered that families of Germans from Russia had moved into our valley, the Clear Creek valley, when they originally emigrated from their homeland. The Clear Creek German-Russians arrived in the United States as part of the last wave of emigrants to flee before the Russian Revolution, this time from the colonies around the Volga River. Finding the farmlands of the Dakotas settled, they pushed on farther west, to Wyoming and southeastern Montana.

Here in the Clear Creek valley, the Germans from Russia settled in a small cluster of homes called the Leiter Estates. At that time, Joseph Leiter owned most of the valley, from his namesake village to the Big Red ranch at Ucross—a total of eighteen miles. Upon inheriting the property, Leiter, whose interests ran more toward farming than traditional cattle and sheep ranching, decided to plant all of his fertile ground to sugar beets. The newly arrived immigrants, destitute but strong-backed, provided large families to do the tedious work of cultivating, harvesting, and topping the beets.

Eventually, Leiter decided to spread out his work force by building small farms throughout the valley. Thereafter, a family would have a home and be responsible for the beet fields on that piece of property. These small tenant farms were eventually sold to the Germans from Russia when Leiter went bankrupt.

I wandered through the old farmstead as my daughters played. Comforting details reassured me of an ethnic connection to these people: the lilac bush near the front door, the tall shady cottonwoods on the west side of the house. Nosily, I pushed open the back door and stepped inside. Tiny bedrooms with nothing more than curtains for closet doors; a square, practical kitchen; a small, plain living room—the floor plan immediately reminded me of my own Grandma Lipp's house.

In the one picture I've seen of Elizabeth Brug, in a local history entitled *The White Root*, by Jean Jardon and Dollie Iberlin, she stands beside her husband, short, dark-haired, strong-faced—not unlike Grandma. I imagine that moving to the United States must have terrified her, yet she set aside her fears, as well as her broken dreams, and began making a home for her family. Did she question God's direction for her life, especially when it took such a brutal, difficult turn? Did she know that she was impacting this new land, this place, simply by her presence? By bearing children who would work the beet farms and change the face of this valley? By planting lilac bushes and yellow roses that would outlive her? By raising her children to remember the old language and the old customs, despite their new lives in a new country? Was she satisfied, having this small, domestic purpose in life?

Actually, her "domestic" impact may still be affecting my family today. Like most rural women, Elizabeth Brug raised a garden. She relied on the waters of Clear Creek to supplement what rainfall her plot received and devised a simple system of trenches between the rows of vegetables. Another trench delivered a trickle of water from the creek to the garden. This small-scale example of flood irrigating allowed her to grow a crop through even the driest of summers. After one particularly miserable year, Leiter's foreman showed him the only patch of green in the entire valley: Brug's garden. Leiter's foreman told him that in northeastern Wyoming, one could not rely on rainfall alone to provide moisture for crop fields.

After viewing Elizabeth Brug's garden, Leiter saw his answer in the waters of Clear Creek. Acting under the recently passed Carey Act, also known as the Federal Desert Land Act of 1894, he designed and lobbied for the creation of a system of ditches to utilize the creek for flood irrigation and for the building of a reservoir, named Healy Dam, to collect and store spring runoff for release downstream all summer long. There is no ranch in the forty miles of the Clear Creek valley today that doesn't feel his—or rather, *her*—impact. Today we use the irrigation waters to raise hay crops to feed our cattle and sheep herds throughout the winters, producing food for millions.

After we returned home from our bike ride, I started fixing lunch. Shawn and I talked about the weather forecast while we fed the kids: there is little promise of rain. Yesterday, like so many summer evenings, ominous-looking clouds gathered over the western hills, carrying a false hope for moisture. In reality, they brought only dry lightning, igniting tinder-dry grass and threatening to destroy hundreds of acres. Shawn was out for hours fighting a blaze on neighboring ground. Now, as he prepares to go back to work, he asks what I will do this afternoon. I consider telling him about an essay idea that has been floating through my mind, but I don't. I'm afraid that if I try to write, the words will not come. So, I say instead, "Oh, probably just fix supper and do some laundry. I need to get some cleaning done, too. Same old, same old." He gives me a slightly sad smile, then steps back outside, shoulders slumping under the white-hot sky.

Elizabeth Brug, who fed and clothed a good share of the work force behind the Leiter farms and worked the beet fields herself, has faded into historical obscurity. Most will never hear of her. To those who do, she was just another immigrant woman, just another wife, just another mother. So were my own grandmothers, my aunts, even my own mother. I long to do more, to make a mark. And yet I fix tuna sandwiches for four kids and my dirty, sweaty husband, sweep up crumbs, irrigate my pathetic garden with water from Leiter's ditch. I long for the words I write to reach far-away souls, for the stories I tell not to fade away. But with each day, each year, each child, taking time to write those stories becomes harder, seems more selfish. My no-nonsense upbringing chides me, and I hear the voice of my grandmothers saying, "There is work to be done, girl! Children to raise, a house to keep! Why waste your time with stories of the past?"

That voice . . . my task-oriented ego loves that voice, measures how I spend my days by that voice. That voice makes my lists, reminds me of chores I've neglected, checks off those tasks I've completed. That voice congratulates me on being a "good" woman because my house stays tidy, my children stay clean, the garden stays weeded. But like the west winds that suck all the moisture from the grass, this voice dries up my joy, my desire, my life.

Lunchtime now fades to mid-afternoon. Waves of fatigue crash over me, and I retreat to the bed, my writing desk dusty, untouched for weeks. I try to tell myself the exhaustion is simply from the endless chores, the occasional nausea from the heat. But I know my body, know that I have to acknowledge the possibility of another pregnancy even if I don't want to. I don't want another child. Grandma Lipp bore nine children; her three daughters have eighteen among them. Fertility seems to be my share of the family heritage, and as I prepare for a fifth child, I see in my future nothing but the never-ending routine of home and childcare that I watched the women in my family live through. Not words, not art, not writing.

As I rest, my oldest daughter, Laura, wanders into my bedroom to ask if she and her sisters can watch television. No harm lies in her simple request, and yet she becomes the unsuspecting target of my buried anger and resentment. Striking without warning, I explode: "No! Why can't you kids leave me alone for a while? Why do you always need something from me?" The unexpected rage shocks her, and she turns away, hurt searing her innocent face.

A shadow of memory flickers across my consciousness, like a cloud that momentarily blocks the sun. I am young, maybe four or five years old, standing in the doorway of my parents' upstairs bedroom. Outside, brilliant light warms the earth, but inside the room is dark and shadowy. The sleeping form on the bed is my mother. I've been admonished not to bother her because she is sick, but still I linger at the doorway, hoping she will rise. The summer my youngest sister is born, I quit hoping that Mom will wake up to play.

When I was a few years older, too old for naps, I sometimes wandered into Mom's bedroom to find her at her sewing machine. An accomplished seamstress, Mom busied herself sewing outfits for my sisters and me, mending Dad's work shirts and gloves, fussing with the satin binding on a baby blanket; she seemed happiest at that machine. Mom's creativity found expression in fabric. My grandmothers and aunts, too, were all accomplished at some form of handiwork: crocheting, sewing, needlepoint, knitting. I bring my sewing machine out of the closet only a few times a year and haven't done needlework since I quit 4-H. Instead, I yearn to spend my afternoons weaving words, sewing up the seams of story, fitting memories together like a patchwork quilt, hoping the resulting essays hold a form and beauty beyond the pieces. But the themes unravel, the stories gap and pull, and my lack of expertise discourages me. Many, many times these past years, I've decided to give up, to spend my days practicing the domestic arts, not literary ones. I tell myself I'm apt to experience more success sewing blouses and baking sweet rolls than I am publishing my essays. And the accomplishments of a domestic life don't risk accusations of neglecting my children, home, and family for more selfish interests.

And yet the simple act of putting words on paper, of seeing blue ink flow out of my pen, refreshes me in a way that domestic pursuits do not. My mother loved to sew, but I don't. Perhaps my grandmothers and aunts satisfied their passions with their tidy homes, flourishing gardens, and embroidered pillows; perhaps not. My instincts tell me they may have pursued their hobbies more because of society's expectations of women than because of any real love.

The menace of drought lies in the fact that it may be ongoing; the sparse summer rains may be accompanied by a lack of winter snows in the mountains. Too many years without snowpack means the creeks will be down, reservoirs like Healy will be depleted, and even Leiter's emerald valley will dry to brown. When I look at the hard, weathered faces of women like Elizabeth Brug and Grandma Lipp, I wonder how many years they lived without snowpack to refill their souls. What dreams did they let go, too afraid of what the world would say about them if they found success?

Today is the third Friday in July. Though still hot, the air temperature has dropped to the lower 90s. We had rain the other night, more than an inch in an awesome thunderstorm, so the grass is a shade greener, the garden a bit perkier. I don't know yet if I'm pregnant, but I do know I'll feel some sadness if I'm not. I've begun to think about this child, to wonder about him, to prepare to welcome her in the way I prepare to write a new story. I've even begun to let myself want this child. Droughts are cyclical, the old-timers say, and even within long stretches of lack, there are periods of blessed relief.

This afternoon, Laura wanders into my bedroom during naptime to find me at my writing desk, rereading old journals. When she asks what I'm doing, I resist the temptation to say, "Oh, nothing," and tell her instead that I am trying to piece together that essay idea. A storyteller herself, she asks more questions, and I try my best to answer them. As she leaves, I dare to pray that I have made a mark on her psyche. If she and her sisters grow up watching me craft stories instead of sweaters, perhaps they will know that women can do more than childcare, housework, and handiwork. Perhaps that will be my impact, my purpose in this world. Perhaps that will be enough.

"C'mon, Sis," I call, following the trail around a bend in the timber. She's off nosing under some deadfall for rabbits and squirrels, so I call again, a bit louder. This time she hears me, picks up her head, and trots back over to the trail. "Good girl," I coo, putting my hand down by my side so she can nose my palm, a habit we've had since she was old enough to reach. Sis has never been the type of dog to jump up to greet me, but she and I do share a language, a bond born of years of walks and hikes like this one. Her ebony coat, normally full and wavy but now shaved for the summer, reveals even more gray hair than last year, and she tends to chase rabbits less than she used to. Then again, I'm sure my head shows a few more gray hairs, too, and I find myself resting more on the rocks around me than I used to.

After one particularly steep climb, I stop again to sit on a boulder and catch my breath. The late July afternoon sun beats down on the open sections of the trail, so I perch on the edge of the timber, just inside the shade line. As my eyes adjust to the change in light, I notice specks of red in the vegetation around the sun-warmed boulders in front of me. Clearly, Sis, who snoozes with her head on her forepaws beside me, couldn't care less, but curiosity tugs at me. Crouching down, I'm amazed to see dozens of red raspberries! I stand and look more closely; there seem to be bushes everywhere on this hillside, and most all of them are loaded with the red fruit. Although I've hunted berries in these mountains for several summers now, never have I encountered more than a few raspberry bushes at a time, and then with only enough fruit for a taste. Usually, my family and I pick chokecherries or gooseberries, both more abundant here.

Stooping down again, I pick a few; the ripe ones fall off their core at the slightest pressure, just the way I remember picking them back in South Dakota as a girl. Their sweet-tart goodness bursts

84

in my mouth—better than candy. By now Sis has noticed that I'm eating—her radar for *that* particular action is as sharp as ever— and she comes over, tail wagging. "No, girl, these aren't for you; let's see if I can find something else." I rummage in my fanny pack for a snack and finally come up with a package of slightly crushed crackers. I'm not as prepared for our hikes these days as I used to be. This afternoon, I decided to go only after Shawn came in from the field early, unexpectedly, and offered to watch the kids so I could get away by myself for a while. So I grabbed a water bottle, my fanny pack with its notebook and pens, and my sunglasses, and loaded Sis into the back of the truck. It was only on the way up the mountain that I realized I'd forgotten to bring something to eat. Now, Sis is satisfied with the broken crackers, and I've got all these berries around me to snack on. Sometimes God comes through in the best ways.

Sis herself was an unexpected gift. We were living on a ranch outside of Boulder, Montana, where Shawn worked after we first married. The ranch produced almost as many dogs as it did cattle, it seemed, and Jessie, the bitch dog, had just had another litter in early spring. Shawn and I already owned one male dog out of Jessie; Jake was loyal, well-behaved, and intelligent around cattle, so we decided to get another male pup. Shawn's dad, however, preferred females, and asked if we would keep one for him until he could come down to visit and pick it up. We named our new pup Pete, but we didn't know what to call the female, since she wasn't ours. Shawn started calling her Sis, as she was Jake and Pete's sister. She and Pete were, of course, too young to work cattle much, so I spent my summer, between graduate studies and visits to my obstetrician, chasing naughty puppies out of the ranch garden and going for long walks near the creek with Sis on a leash because she never wanted to follow along. But when I would sit on the front porch that summer, either dreaming of what our first child would be like or mourning my dad, who had died that spring, Sis would climb into my ever-smaller lap, lick my arms, and go to sleep. By the time Shawn's parents came the next fall to meet their new granddaughter and pick up their dog, I couldn't let go.

"Hey, girl, let's walk a little farther and see what we find." The raspberries have given me a new drive to explore more, hunt farther on up the trail. I don't feel the strain in my legs as much now; the pleasure of spotting another bush outweighs the heat and fatigue of this afternoon.

The entire area seems to be covered in thickets; I've heard of places like this but have never seen one. The thrill of finding the fruit takes me back years, to my Grandpa Quinn's garden, where picking raspberries was one of the chores that fell to my sisters and me. Grandpa's bushes were domestic, and so thick and tangled that only skinny kids could fit into some places to pick the sweet red jewels.

Grandpa's garden was a wondrous place by central South Dakota standards. Of course, the usual practical crops of carrots, lettuce, tomatoes, and other vegetables grew there. But around the edges of the garden were plants you didn't usually find in a place where rainfall was erratic and land was apportioned to plants that were most certain to produce a crop. In early summer, we girls crawled through the dense mat of strawberries, picking gallons for Mom and Grandma to put up as jam. In the fall, if we'd had a wet summer, we sometimes had enough luck to pick a crop of apples from the trees planted on the west end of the garden. And always in July, there were the raspberries from the thicket that covered nearly a quarter of the plot.

This garden, however, was Grandpa's retirement project. When he and Grandma had lived at their first South Dakota farm, just a mile down the road, she'd tended to the gardening herself and had grown the crops there that were the most practical and economical. Grandpa had been too busy milking cows and raising field crops to fuss with such work. But when he turned over management of the dairy to my mom's oldest brother, and he and Grandma moved down the road, the planting and tending became one of Grandpa's favorite pastimes. Though he still fed out calves, drove the tractor, and helped with whatever farm work was needed, he'd find time to lean over the garden fence at the end of the day.

Sometimes a relative or neighbor would drive into his yard to find him in the garden, flat on his belly. "Dad, what are you doing?" my mom would always ask, slightly alarmed.

"Oh, this darn jenny," he'd spit, pointing to a skinny vine that curled around the tops of the carrots. Besides farming, Grandpa had worked for years for the Soil Conservation Service and bore an extreme dislike for weeds of any kind. One in particular—field bindweed, or creeping jenny—became enemy number one. Rather than spray with herbicide, he would patiently, painstakingly dig away at the ground with a long screwdriver, careful not to chop the roots, until he could tug the entire plant from the ground. He piled up the unearthed weeds, then carried them out of the garden to the burning barrel. Despite his valiant battles, the jenny seemed to be winning the war, coming back each year without fail.

Sis and I have stopped to rest near the creek; she lies down in the stream and laps water from between her front paws as I empty my water bottle. I think we've come to the end of the thicket; here deeper shade and cooler temperatures have made the berries scarce for the last hundred yards. I've decided to pick what I can on the way back down; I'll use my water bottle to hold them. Not a practical decision, I'm sure; survival experts would tell me not to drink the last of my water before starting back down the mountain—what if I fell and needed to wait for a rescue? But practicality isn't my strong suit; I just can't let Shawn and the kids miss out on the taste of these berries. Besides, I've got Sis with me.

I remember arguing with Shawn for several weeks when the time for Sis to leave approached. Naturally, he didn't want to disappoint his parents, who were expecting a dog to take home. His main objection, however, was that having a female was not practical, for obvious reasons. "We'll have to get her spayed, otherwise I won't be able to use her for any work. She'll be just like Jessie, always dragging a bunch of pups."

"So we'll get her spayed—so what? I don't want her to go with your parents."

Shawn's folks, too, tried to convince me how impractical it would be to have a female on a ranch full of males. But my emotions—or my maternal hormones—won out. Finding the money to have her spayed, however, was not as easy; our first daughter was quickly followed by a second. So, Sis did mother two litters of pups before we finally got her to the vet. It's an act of negligence I don't really regret. Somehow, having those pups made Sis calmer, more willing to stay at home when I needed her to. She also became gentler around my own kids, allowing them to pull at her black curly hair or sit on her back. Most importantly, she became more protective of both my kids and me. More than once she's snapped at the heels of a ranch worker who moved too quickly towards my girls.

As Sis and I mosey our way back down the mountain, picking as many berries as will fill the bottle, I begin telling myself stories, talking out loud to the bushes and sky. It's a habit Sis has grown accustomed to: occasionally she'll cock her head quizzically, as if to say, "Were you talking to me?" Laura, who's six months younger than Sis, used to say, "Mom, are you talking with yourself again?" Mostly, though, the creatures who've spent the most time around me know that I talk to myself regularly. It's more correct, however, to say that I am talking to someone who happens to be present only in my head. Today, I begin telling my kids—who are actually at home thirty miles away—about Grandpa Quinn. I tell them stories I don't want them to miss: about Grandpa encountering a wolf as a young boy; about him being born in a log cabin; about his attempt at college as a young man, an attempt thwarted by the Great Depression. I tell the kids stories about him because the Grandpa Quinn they now know is only a shadow of the man I once knew.

Grandpa Quinn, of all my ancestors, endowed me with a storyteller's voice. Grandpa was famous in three counties for his stories, and his travels for work or pleasure always took him to homes where he would love to visit for hours. He would preface a story by saying, "Maybe I've told you this before, but . . . ," and the listener became his captive audience. Yes, we kids did hear many of those stories over and over again. Yes, perhaps we could have used our time more productively than to sit listening to those words we already knew. But in truth, I remember the hours I spent listening to Grandpa's stories as some of the most precious of my childhood, and I wish I could give my own kids that same kind of privilege. There was always plenty of work to do around the farm and garden, but Grandpa must have believed that the stories he told were important for people to hear. He gave time to storytelling, the same way he gave time to digging weeds and planting impractical plants. Grandpa knew how to give time to life.

Unfortunately, time no longer seems to be on Grandpa's side, and his storyteller's voice is silent now. Grandpa's decline began shortly after Grandma Quinn passed away, nearly eight years ago.

As so many widowed people do, he lost interest in the pursuits that had once been his passions. Gradually, he also lost the ability to care for himself. My own mom stepped in, stopping every day after her job at the school in town to wash and clean, cook his supper, and keep him company until the early evening. She even baked him a cake every week, as Grandma had done. But during the long days when she was at work, he was alone. He perked up for a while when he got a dog, Buddy; the companionship comforted him. Others besides my mom took their turns staying with him, and this company also became a lifeline to him. But as the years progressed, caring for Grandpa became a task that one person simply could not handle, no matter how dedicated. He was moved to the Alzheimer's ward of a nursing home in the Black Hills, near two of his sons, but hours away from his farm and the life he had known. Grandpa's oldest son moved one of his own kids into Grandpa's house and assumed control of the entire farm. By the next summer, the raspberry bushes had been plowed under.

Grandpa now spends long hours in the nursing home, unsure of his surroundings. He asks frequently to be taken "home," back to Timber Lake and the farm. He talks of being ready to die, and of missing Grandma. He doesn't tell stories, or plant gardens, or even enjoy sweets. When I go to visit, he squeezes my hand and smiles but admits that he's not too sure who I am. When my mom makes the three-hour drive out to see him, he sometimes doesn't even recognize her.

Sis noses my palm again, and I reach down to stroke under her chin. "Hey, girl," I whisper, looking into her deep brown eyes. I swear, sometimes those eyes seem almost human in their intelligence. Despite her having helped Shawn work hundreds of cattle and sheep over the years, in my mind Sis is my dog, my companion. What would it feel like not to recognize her, or not to be recognized by her? I have four daughters of my own; how could I possibly not know who one of them was? How terrible it must be for Grandpa to be told, "I'm your daughter, Dad," and not remember. How much more terrible to be the daughter or son who says, "Hey, it's me; don't you know me?"

We don't know much about the emotional connections among animals, but I'm sure they feel them. From listening to cows bawl for their babies at weaning time each fall, to watching Sis nose around for her pups after we'd given them away, I know animal mothers suffer from being separated from their babies. And although Sis was too young to react when her brother Pete was run over as a pup, she moped around lethargically for days when, years later, we lost her older brother and working partner, Jake. Since Jake died, we've had other dogs, including Max, the young male who waits for us at home. Sis has patiently trained him to work livestock and indulges him with boisterous wrestling matches on the front lawn. Their exuberant play cannot be described as anything less than joyful.

Sis rolls onto her back, wanting her belly scratched. Satisfying her need quells the rising sadness in me, and I am overcome with wonder at Sis's pure, simple love, and at this gift of a perfect afternoon. So many blessings fill my life: these glorious mountains with their amazing crop of raspberries; my pets; my children waiting at home, including our delightful baby, Emily, whose development had induced such a depression in me last summer. I could have spent the day at home writing or working in my garden instead of driving all the way here to hike—I could even have hiked the hills surrounding our home. Instead of stopping to pick and eat raspberries every few steps, I could have passed them by and gotten much more of a workout. I could have chosen to live a life with fewer pets, fewer kids, and more time for a career, money, and worldly achievements. But to do so would be to miss out.

I'm not a doctor and certainly don't know much about the disease that has stolen my Grandpa's voice, his memories, his spirit. However, I do have my own theory about Alzheimer's disease: from my own observations, I think the depression that often follows some traumatic event opens the way for the downward spiral of Alzheimer's, in which those afflicted gradually lose their ability to feel wonder, or awe, at whatever life remains after their loss. After Grandma's death, Grandpa didn't want to garden anymore, because that had been their shared pastime. He no longer told stories, because she was a main character in most of them. Perhaps he even stopped recognizing his children because they reminded him so much of her. In attempting to insulate himself from the paralyzing sadness of her death, he put aside all the treasures he'd enjoyed during her life, and left himself no possibility for joy. By contrast, my mom, who still grieves for my dad, manages to feel delight in her grandchildren, her dog, her students. Whether due to her age or her gender, Mom has survived the loss of her spouse more successfully than her father has.

I know I won't have Sis forever; already, as I've said, she is showing the age that will take her from me. I know, with each child I bear, that I take a risk of a heartbreaking loss, the same loss that Grandpa suffered when his first daughter was born prematurely. I know that even impenetrable raspberry thickets can be plowed under. Life would be simpler if we didn't love at all, if we protected ourselves from heartbreak by not allowing ourselves to feel.

Sis and I reach the trailhead and load into the pickup again. At home, we share one last hug before I burst into the house, eager to share my water bottle of wild raspberries with Shawn and our five kids. As I look outside, I notice that Sis is limping slightly; there probably won't be many more hikes like this afternoon's. I know I will weep for days when she dies. But as I remember back to that Boulder ranch, to afternoons spent walking by the river and lounging on the front step, I know that, along with my first child and my Grandpa's raspberry bushes, one of my life's greatest gifts was growing that summer—a spare puppy who would be a friend with whom to share my walks, my children, and my life.

God, this room is cold, I think, as nurses strap my arms to supports that run perpendicular to the operating table. They insert IVs, expose my swollen stomach, prep my abdomen for the incision. The image of Christ, arms outstretched in crucifixion, floats into my fuzzy brain; an incongruous idea, to think of death now, in the midst of all this preparation for life. And yet, a death is occurring—part of my life ends today. Bright, hard lights stare down at me, unblinking, unfriendly, so I look over at Shawn. He takes my hand and smiles. "Some date, huh?" he jokes. We're both relatively calm—at least more so than we were eleven years ago, the last time we faced a caesarean section.

Today our last child will be born. We've known the date of this delivery for months, ever since being informed that our small rural hospital was no longer allowing women to deliver a child vaginally after once having had a caesarean delivery. The irony surprised and angered us: after our oldest, Laura, was delivered surgically, my doctors encouraged me to try vaginal births with each of our succeeding four children. Even Emily, delivered only seventeen months ago, in this same hospital, by the same kind-eyed doctor who now smiles at me from behind his surgical mask, was a vaginal birth. Yet, due to rising liability costs and a shortage of anesthesiologists in this area, the hospital board has since determined that women like me can no longer attempt such births.

And so, several months ago, Shawn and I sat in our doctor's office looking at a calendar. Based on my due date, he offered us our choice of two successive Friday mornings in June; we chose the one that conflicted less with our second daughter Carmen's birthday. An odd, somewhat unnatural feeling settled over our lives. "Hard to believe yesterday I was taking the kids to story hour," I say to him now. "Some of the other moms were talking about their plans for the weekend, and it was so weird to say I was having a baby today—

like I was getting a haircut or something." He just laughs, but his smile doesn't reach his eyes. He's sad, too, I think.

Just as we've known today will be the day, we also know that this child will be the last. After delivering this baby, my doctor will perform a simple, quick procedure to cauterize my fallopian tubes, and I will be sterile. Sterile. What an awful, cold word . . . just like this room. Dr. Scott has already told me that he'll ask one last time, just before the procedure, if I'm sure. How will I be able to say yes? I don't even think I want this procedure anymore. In eleven years, I've been pregnant seven times and delivered six—well, just about six—children. How do I say no more?

Fertility is the quality most revered and desired in agriculture. Sterile cows are culled; barren old mares are sold. Without fertility, the industry doesn't survive. Males of most species are regularly castrated, sterilized to make meat, to prevent wandering, to curb fighting. But to sterilize a female used to be a rare process, reserved only for pet dogs and cats. Recently, ranchers have begun spaying some of their heifers to sell for meat, but even this development is relatively new. Traditionally, heifers were kept intact to breed and reproduce, thus continuing the cycle. Fertility wasn't just a good thing; it was the good thing.

And so it is with Shawn and me. We manage a ranch: the cows, ewes, and fields must be fertile for us to have a paycheck. On the side, we raise Quarter Horses, breeding our stud to a band of mares, judging each female on her potential for conceiving and carrying a foal. Even in my teaching job, I rely on other people's fertility to keep students in my room—the reason teachers often jokingly refer to big families as "job security." If fertility then, is so crucial to our lifestyle, why are we choosing to terminate our own?

Strangely, my first thoughts of making this decision arose from a discussion with other teachers. Five of us, all women, compose the high school English department in our district; one afternoon, after an intensive day of reading and assessing 120 sophomores' writing samples, we began visiting about my upcoming delivery. I was still complaining about having to undergo the c-section, when one of my friends and colleagues said, half-teasingly, "Well, Darcy, maybe you should just have your tubes tied while you're on the table. That way you wouldn't have any more kids."

Her suggestion offended me a bit at first. . . . What's wrong with having lots of kids? I remember thinking. Sure, my last two pregnancies were unplanned, but not unwanted. I went through a depression when I discovered I was expecting my fifth child, but Emily, the daughter I delivered, brings pure joy to my life now. With this pregnancy, I lived in denial for months, not admitting my condition even to myself until I saw the baby move in my belly one night as I lay in the bathtub. But then I accepted it; I wasn't even depressed this time, because I was back to teaching and writing, feeling stronger and more capable than I ever had while pregnant with Emily. If my role in life is to have lots of kids, I can handle that. There are worse things.

Once my defensiveness went away, I was more shocked than offended at Karen's suggestion. Choosing to end the fertile period of my life—a phase of more than twenty years—could I do that? The possibility overwhelmed me. Besides being proud of my family, I firmly believe that fertility is a gift from God. Sisters and friends have struggled through heartbreaking procedures in order to conceive; I know just a little of their trials from the four years we tried unsuccessfully for our fourth child. I also know the pain of losing a baby; each day I thank God for the ones on earth, the children I can see and hold and kiss. What kind of person says "No, thank you" to gifts like that? What kind of sinner chooses to end their fertility? And is it that far a step from my choice, my thinking, to the choice of a woman who ends an actual pregnancy?

Six weeks after the birth of our first child, Laura, before we were ready to have a second baby, I went back on birth-control pills. However, because I was nursing, the doctor put me on a different type of pill, one that, instead of preventing conception by suppressing hormones, worked by preventing a fertilized egg from implanting in the uterine wall. I thought about that difference during all those long afternoons spent holding my baby and recovering from my first surgery. Gradually I realized the implication: life begins at conception, so it was possible that a baby was being conceived, but then dying within me. I might have been responsible for another human being's death already. As soon as I realized what we were doing, I tearfully told Shawn I couldn't take the pills anymore, and once he understood my point, he agreed.

They're getting ready to begin now. A second doctor has joined Dr. Scott, and more nurses have entered the room. Forget about my earlier calm; it's gone now. I know what can go wrong. I've been through so much more since that time eleven years ago when, so tired after thirty-six hours of unproductive labor, I agreed to the doctor's decision that it was time for surgery. I'm just not the same girl who lay on that operating table then, nor is Shawn the same boy. We're parents of five other children now, who wait with our good friends for news about whether they have a brother or sister. We've lost a child, lost homes, pets and loved ones, and nearly lost each other. Back then, we naively believed that life would simply go our way. Now, each child, each move—sometimes each day— requires an act of faith; we know how lucky we've been. And we know that it could all change within the next hour.

Once I suggested the sterilization to Shawn, I discovered that he'd already been thinking about it but was unsure how to bring it up. "It doesn't seem fair, asking you to go through that when you already go through so much," he said. He had been thinking that it was time for him to have a vasectomy. The procedure itself—or who became sterile—didn't matter much. What mattered was that we were both thinking about ending our child-bearing years. It was one thing when we'd postponed conception, either before Laura or between any of the children; it was another to cut off the possibility of ever conceiving again altogether. So much of our identity as a couple—so much of my identity as a woman—was tied up in having children. What would happen to us when that part of our life ended?

This caesarean section is part of Shawn's reason for wanting one of us to be sterilized. How frightening this must be for him, watching them cut me open like this. His eyes are teary and worried, and his grip tightens. "I love you," he says, over and over. I try to reassure him that I'll be fine, but there aren't words to take away his fear. Despite the advances of modern medicine, he knows that I could die, or the baby could die. He also knows that if we continue to get pregnant I'll have to have this operation every time I deliver. He wants to be done because he wants to protect what we already have. I close my eyes to pray, wondering if God will listen to me today. Don't punish this baby, I pray. Don't punish Shawn or my other children, either. They need me; don't take me away from them. I'm sorry, I'm sorry, I'm sorry.

We haven't told anyone else about this decision; in fact, to me it even feels wrong to discuss it with the medical staff. I know my up-bringing, my Catholic-girl guilt, causes me to feel so embarrassed. According to my religion, women are not supposed to use birth control, have abortions, or choose operations like this one. I find myself, yet again, caught between the rules of the religion I love and the realities of my life.

It's amazing how many people feel they have a right to comment on family size when they see a lot of kids. I wonder if people with just one or two children receive the same types of rude remarks. I've had total strangers approach me in the store, asking incredu-lously, "Are they all yours?" We've been subjected to old cowboys in cafés telling us—with three daughters, at that time—stories about so-and-so, who kept on having kids until they "finally got their boy." Even our families have felt the need to offer family-planning advice, ranging from my nephew's "Another one? You're kidding!" to the famous "That's about enough now" that Shawn's mother has admonished us with after every birth since Maria, the third.

So, during this pregnancy, when people who have no right to ask have asked anyway, I've been tempted to respond to their "Are you done yet?" with something just as rude, just as inappropriate. Instead, I've joked that we have to be done, as we've run out of seats in the car. Just two weeks ago, we traded in both our seven-seat minivan and our crew-cab pickup truck for an eight-passen-ger SUV that could pull a horse trailer.

And that is the reality of my life, whether my religion can ac-commodate it or not. Feeding, clothing, sheltering, medicating, and transporting six children requires a lot more work than two or three. And I want to do so much more than just give my kids the basics; I want to provide vacations, enrichments, college. I want to be able to listen to them, support them, give them individual atten-tion. I never want one of my children to feel I'm just too busy for them. I love each of my children bone-deeply, heart-wrenchingly. To have any one of them feel that I don't love them as much as an-other, or that I'm too preoccupied with the details of raising them all . . . wouldn't that be a sin greater than undergoing sterilization?

I think of Laura, our first, until today our only c-section baby. When we announced our fourth pregnancy, when she was six, she couldn't wait to tell her friends, her teacher, everyone she knew. When, not even a year later, we told her and her sisters about the fifth baby, she was scared, because our family had suffered through the ordeal of Cody's premature birth, and she was afraid we would go through that again. But after we assured her that this delivery would be fine, that we were watching carefully for signs of trouble, she was excited again. When, last winter, we asked her what she thought of having another baby, she seemed tired and more than slightly disgusted with her parents. I wonder if she's ever felt forgotten among all these babies, or felt like she's just a convenient babysitter now.

"OK, Darcy, we're about to start." It's Dr. Scott, warning me that he will be making the incision now. What an eerie feeling, to realize that your body is being cut open, but not to feel it. I can hear the buzzing sound of the equipment, feel a sensation of pressure on my pelvis, but nothing more.

"Don't look," I whisper to Shawn.

"Don't worry," he whispers back. "I'm only looking at you from the waist up this time." Eleven years ago, he thought he was brave enough to stand up and peek over the curtain that veiled my abdomen. After all, he'd seen dozens of caesarean sections performed on cattle and had even performed many of them himself. Blood and guts he knew about. But when it was my blood and guts he saw . . . I remember him sitting back on his stool quickly, too quickly, and the nurses watching him carefully for a faint. Now, there is no such show of bravado. He's here for me, as he's been through each delivery. For me and our baby.

Perhaps this decision has been a bit easier for Shawn because he is more practical-minded. He sees the implications of more children on the kids we already have. We're both working, both exhausted at night; he knows that these children don't need parents who are always short-tempered and worried about money. "We have all we can handle, Darc," he's said so often these past few months.

Or, perhaps it's been easier because he makes life-changing decisions like this often. I've only ever chosen to sterilize dogs and cats, particularly my female dog, Sis. He chooses whether to spay the heifer or keep her for replacement. He chooses whether to take the cow in for the c-section or try to pull the calf. Sometimes, he even has to choose between saving the life of the mother cow or the life of the calf. He chooses the mother.

Suddenly, I can feel Dr. Scott tugging at my insides, separating the uterine wall to get the baby out. I don't remember Laura's birth being this way. Actually, I don't remember feeling anything when she was delivered. This time, although I don't feel pain, I feel a distressing jerking and pulling at my insides. I grip Shawn's hand harder, my fingernails digging into his skin, and take a deep breath. This definitely isn't what I expected. Each time the doctor pulls, I can feel the tension, feel my muscles and tissues resisting his efforts.

It seems to go on interminably. I thought Laura had been out in a matter of seconds; this delivery doesn't seem to be progressing at all. Is something wrong? I try to read Dr. Scott's expression, but his eyes show me nothing. The jerking, pulling, tugging continue. Just make it stop, I pray. Help me through this. There comes a point in any delivery, or at least in any of my deliveries, when the woman realizes that no one can truly help her. The doctors and nurses will exert all their professional skills; the husband, if he's there, will offer all the emotional support of which he's capable. But the process of delivery is up to the woman, to the woman's body, alone.

During my first four deliveries, I was aware of this realization only fuzzily, my thinking obscured by pain-relieving drugs that had been injected into my spine. Those times, the loneliness of realizing that nobody could do the job for me registered only as weepiness, helplessness. But last time, with Emily's delivery, I was not able to have any anaesthesia. To compound the problem, she was not turned correctly, and I had to do long minutes of pelvic rocking to adjust her as she entered the birth canal. The pain was back-breaking; it felt as though the lower half of my body would split in two. No wonder cattle bawl and roll their eyes; no wonder women scream. At one point in that long, exhausting labor, I buried my head in the pillow and cried in despair. There is no one to help me here. Jesus, please, You have to help me. I can't do this anymore.

Now, I've come to that same point, even though I don't feel pain. The pulling is almost more frightening, however. Is my body resisting? What if this baby isn't ready to come? It's so insane, this scheduling of a delivery, this scheduling of even the most miraculous parts of our lives. When did humans decide they could play God?

And then it is over. "It's a girl!" Dr. Scott cries as our baby squalls. A girl. Katie Lee. Our fifth daughter. I look at Shawn quickly, trying to catch his expression before he can cover it up with a smile. Is he disappointed? He'll only have one son now. Does this matter? Does it change things? But he's looking straight back at me, grinning all the way up to his eyes, not hiding a thing.

Today is the day after Katie's birth. Already, of course, all of her siblings have been here to meet her. In addition, I've had visits from three other good friends, and now Karen, the teacher-friend whose suggestion sparked so much, is here to meet the baby. I did play God yesterday; I said yes when Dr. Scott asked. I think about all the other "yeses" I've said in my life: joyful yeses to scholarships and awards, to Shawn, to each of our incredible children. This yes was not a joyful one. It was a sober, soft, sad "yes"; but I hope, also a "yes" to more time for the children I have, to the chance to raise them carefully and conscientiously so that they will grow up to be six more chances to save the world.

As Karen holds Katie and we visit, there is a knock on the door. "Come in," I call, really not knowing who else to expect. A priest enters. Oh, God. Do they know what I've done? The ridiculousness of my thinking chastises me, and I greet him as politely as I can.

"You said on your hospital check-in sheet that you are Catholic," he states after introducing himself. I already know who he is; he heard my children's Confessions once, when we were visiting at the church here. They liked him, said he was kind to them. "I was wondering if you wanted to receive the Eucharist."

This is a test, I think. It must be. This is where I'm supposed to say that I'm unclean, that I'm a sinner not worthy of receiving Jesus. I struggle, honestly not knowing what to do. Karen is not Catholic; she cannot help me. The rules say that I'm in a state of mortal sin, that I'm supposed to go to Confession before I receive the Eucharist. The kindly priest smiles at me; what would he say if I told him the truth? Would he smile, still? The seconds seem to stretch forever as I decide what to do. God already knows what I've done, how sorry I am, how I wish it could have been different. He knows my heart, doesn't He?

Jesus died to redeem us from our sins. He died to save us from all thoughts that we are in control. Maybe, instead of being a test, this is a gift. Maybe I should just take the gift. And so I say yes. Again. A grateful, contrite, profound yes. I'm so sorry, Lord, if I've made a mistake. Please forgive me; I'm just so sorry. Please help me move forward. And thank you for my children.

Then I receive the ultimate gift from the religion that both loves and restricts me. My friend watches reverently, holding my baby. I think of all the women in the world who can't have babies; think of all the poor creatures whose lives end after they're done having babies; think of all the babies of the world whose mothers didn't want them; think of all the women who have, for whatever reason, said no to a miracle. I know I am so lucky. And in my gratitude, my relief, I do the only thing I can think of: I thank the priest, take Katie from my friend, and nurse my daughter, my last, blessed child.

The balloon-full moon shines down on the countryside around me as I speed down the highway, straddling the centerline with my oversized SUV. I focus on the point where the light from my headlights meets the semi-darkness, giving myself at least a little time to slow down for deer crossing the highway. Driving down the center of the road helps, too; by this time of year most of the deer are grazing in the barrow pits—in South Dakota we would say ditches—so I'm putting an extra cushion between myself and them by staying in the middle of the highway. The glow of an oncoming vehicle's lights warns me when I need to pull back into my lane. Not a legal driving move, but a functional one, at least. It's a trick, a dance just on the edge of safety, that I learned from my husband—one of many that have helped me adapt to life in this part of the West.

Tonight, though, I'm reluctant to admit to anything good coming from sharing my life with my husband, though, of course, there are six sleeping children in the ranch house miles behind me. I checked them all before I grabbed my purse and slammed the door on my way out of the house; Shawn was already in bed, ending a long night of arguing about the lifestyle we're leading—or rather, leaving.

After returning to ranching more than five years ago, Shawn recently announced that he is ready to leave this career again. The rancher we're currently working for has changed his mind about management practices, employee benefits, and other issues on the ranch, to a point that has angered and insulted Shawn. We've discussed, cried, and argued for weeks about which option to take: remain on the ranch and put up with the changes, or find Shawn a different job and move to town. In the meantime, Shawn's found a job in the methane fields and a house for us to rent; today he gave his notice to his employer. Although I knew this day would come,

I still wasn't ready, and tonight I am angry beyond words. Somehow, I could go along with his decision in theory. The reality of it, however, enrages me.

I pull into the parking lot of a local bar on the edge of town, already hearing the music from inside as I turn off the engine. I don't know what I'm doing here; other than a few outings with girl-friends or colleagues, I haven't entered a bar without Shawn by my side in fifteen years. But something pulls me toward the doors; I pocket my keys, my identification, and some cash and walk in. At the bar, I order a Diet Pepsi. I rarely drink anymore, and besides, I'm still nursing Katie. I know I didn't come here to drink, because that would only cloud my thinking further. Something else drew me here.

I'm angry not simply because Shawn has given his notice, but because the awful familiarity of my position makes me disgusted with myself. When Shawn and I married, I little realized how my life would revolve around his career. Whenever he has left a ranch, for whatever reason, I've been forced to leave a home and start over again. It all sounds so 1950s: the wife packing up boxes and leaving behind memories because the husband isn't happy. I'm an educated, talented woman with six children, a job at the high school, respect from colleagues and students . . . and yet, in a few weeks I'll be slogging through the mess of moving, drying the tears of my children, hoping that all will work out again. Shawn right-fully feels insulted by the changes in management at the ranch, but I feel even more insulted by the lack of options I've left myself.

Of course, I do have options, one of which is to dig in my heels, refuse to move, and force Shawn to work at a job he resents. I could play the bitch; tonight it would be easy. But how long could I keep it up? And at what price to my marriage, my kids, myself?

The band resumes playing after its break, and I turn on my stool to watch. A cowboy comes up to my elbow to ask if I'd like to dance, but I turn him down; I still dance with only one man. The rhythm of the music and movements of the dancers calm me, and soon I forget my anger as I watch the people.

Watching the floor of a western bar on a Saturday night is a pastime Shawn and I have shared for years. Although we love to dance, sometimes watching is more fun. Tonight I see several young couples two-stepping around the floor; some of them move well, and some stumble clumsily along, gamely trying to imitate the steps of the more expert dancers. A group of young women, probably just over the drinking age, gyrates in a circle near the center of the floor. To my left, a couple dressed in torn jeans and Harley-Davidson T-shirts, bodies and lips pressed together, sways, barely moving, to the upbeat country song. And around the edge of the floor, one obviously married couple, probably in their mid-fifties, dressed in nice, though not fancy, clothes, skims easily over the sawdust in moves practiced over a lifetime.

The woman draws my attention, so I focus more closely. She's wearing a rose-colored dress with a skirt just full enough to swirl as she turns—a dancing dress. Strapped to her feet are high-heeled shoes; the last time I danced in heels was at my wedding reception, and even then I ended up barefoot after the first three songs. She, however, glides expertly. The band shifts to a three-count song, a waltz. I giggle to myself as most of the couples on the floor try to adjust their two-step to this new beat, but she and her husband easily begin the traditional step. She's been dancing with him all her life; I can tell by the way she reaches for his hand after the turns, naturally anticipating where it will be. I watch them through several songs, having lost interest in the other couples. Her husband rarely steps backwards in the moves, almost always propelling her across the floor. She follows his lead, swirling, turning, dancing backwards with elegance and grace. Although her husband does move well, *she* simply shines.

God, I'm just so tired of dancing backwards. Watching this long-married couple returns my thoughts to my own relationship. In fairness, I must admit that Shawn did leave Montana, his home state, to move to Wyoming when I accepted a challenging job in Dubois. He also shouldered our family's entire financial burden when I stayed home with the kids for four years, just after we moved to Buffalo, Wyoming. Through most of our marriage, though, his career choice has determined both our home and our lifestyle. I've been spinning to his turns for so many years that, tonight, I'm exhausted.

But as I watch the woman in the dress, I realize that her beauty and grace come just as much from being able to follow her husband's lead as from any talent of her own. She's probably just an average ranch wife; tonight is probably her anniversary or some other special occasion. Perhaps she learned to dance as a young girl, practicing with her father in the living room, maybe even standing on the tops of his feet the way my daughters do with their dad. Perhaps when she first met her husband, she wasn't able to follow his lead so well; she stumbled, possibly, or tried to lead herself. Somewhere along the way, however, she learned to dance backward, and now the two of them dance as one, surpassing the abilities of any other couple on the floor tonight.

In the middle of a turn near my bar stool, the woman catches my eye. I've danced enough to know that she may not even be focusing on my face, but still, she smiles. In that moment, I feel like she is smiling at me, blessing me. I feel like she knows. And inside my head, a voice says, "It's OK, Darcy. Just go back home for now."

I kneel next to my flower bed on this drizzly October afternoon, saying goodbye. Here near the faucet are the remains of this summer's four-o'-clocks; they were the only flowers to grow in this hard clay bed our first summer here, and they've reseeded themselves every year since. Here in front of the window to Laura's room are the three tiger lilies I planted to remind me of my Grandma Lipp's tiger lilies back home. Over there is the yarrow, planted on the advice of my good friend and neighbor Annette, who's also struggled to grow flowers in this red clay. And here, sheltered by the steps to the deck, still green, is the blue flax I planted to remind myself of my friend and mentor, Linda, at whose home I attended a writers' retreat many summers ago.

Four and a half years ago, when we left another ranch and moved to this place, Linda sent me a postcard. After talk of various writing opportunities, she ended by saying, simply, "Keep in touch, and keep adapting—it's a survival skill for sure." Linda ought to know: her life and her books have been filled with stories of adapting to loss. But still she writes and recently has turned to helping other women learn to write, even editing anthologies that give women opportunities to share their voices.

I turn again to look at the bed. No frail roses or prissy peonies here—these less-showy flowers adapt to this soil and climate through deep roots and thick, tough stems. All but the flax. Its thin, thready stems seem like they'd be the first to succumb to the extremes of wind and temperature of the West. Instead, the blue flax growing in my flower bed only echoes the wild plants that flourish untended in the barrow pits and meadows throughout this region. Each lavender-blue bloom lasts just a day, petals littering the ground by evening, but the plant itself produces color and beauty from May through July each year. The seemingly fragile plant survives by letting go of its blooms to expend its energy on new ones. The blue flax dances between life and death, between holding on and letting go.

Of either of the sexes, women are more often forced by nature to let go of the old to make way for the new. We let go of our girlhood to become women, let go of our bodies to become mothers, let go of our youthful beauty to become elegant, graceful, wise old ladies. We let go because we are strong enough to do it. Because we have to. Because we know. We know that holding on too tightly will cause us to wither and die.

Letting go of this place tears at my heart. I've planted gardens and flowers here, buried old horses and baby foals here, conceived and borne three children here. In five years, a place can become such a part of your soul that leaving it feels like leaving your self. More than the place, though, I mourn the lifestyle we're leaving. I mourn the lost opportunities for my children to ride their horses and tend bum lambs. I grieve for the look of satisfaction on Shawn's face, a look I know I won't see when he comes home from the methane fields. I cry for the loss of space, of wildlife, of a life based on the simple cycles of birth and death, food and water. The tears that have pricked at my eyes all afternoon now spill over. Even before I leave, I am yearning to come back. I don't necessarily mean to this *place*, although I would love to return; I need to come back to this *life*.

And yet, deep inside I know I must let go. And so I pray. I pray that another ranch wife might find her days brightened by these beautiful flowers. I pray that this ranch and this lifestyle will survive, even if I'm not to be a part of it. I pray that Shawn and I can move past our mutual anger in order to be strong for our children. And most of all, I pray that I will have the grace to dance backward, in order to move forward.

Early Sunday morning, July 1996. The icy clear water of Gannet Peak's runoff splashes down the rocks of Torrey Creek, just below my feet. I perch on a boulder, sun-warmed just enough to be comfortable, and count the species of wildflowers in the clearing around me: Indian paintbrush, lupine, fairy slipper . . . God's palette sparkles with brilliant orange, green, and violet, and the sunlight intensifies each hue. I breathe in the pure air, the scent of cedar and mossy rock, of tall pine and bubbling creek. *Here,* I think, *here I truly feel close to God.*

But then guilt pricks me. It *is* Sunday morning, after all; shouldn't I be in church? In an instant, my chagrin shatters my tranquility, and I wrestle my demons. My head and heart know that to worship God requires no specific location, no predetermined ritual. Yet a part of my psyche, embedded as firmly as my identity, holds up my family's standard of religious observance, and I know I fall short once again.

In my childhood and even into my early twenties, attending church on weekends was as natural a part of my life as going to school on Monday mornings. It was simply what we did, my people and I—simply who we were. Rules for life were simple: Work hard. Don't spend what you don't have. Don't swing your Rosary beads. Marry another Catholic. And always, always, attend Mass on Sunday.

Catholicism was not simply my family's religion; it was our lifestyle. Besides Mass every Sunday, elements of our faith intertwined with our daily life. The crucifix guarded us as we slept and watched TV; a painting of the Lord's Supper watched over our own meals. In fall, we attended the Catholic grade school in town, taught by the same nuns year after year. In winter, we played with the figures from the crèche nestled under our Christmas tree. Spring brought First Communions, May altars, and Rogation Days when the fields received special blessings. During the summer, we prayed the Rosary during thunderstorms and sprinkled holy water around our house whenever tornados threatened.

And summer was when Sister came to visit. Sister Margaret Francis, our Grandma Quinn's sister, had joined the convent and lived her life as a nun since the age of sixteen. Each summer, she returned from Iowa to Grandma and Grandpa's farm for an extended stay. She said no place on earth was as beautiful as the plains of South Dakota. And she seemed to know: her teaching and missionary work had led her all over the United States and to several foreign countries—farther than anyone I knew had traveled.

Life sparkled around Sister. She loved to hear our stories of school and friends and peppered us with intelligent questions about whatever we were learning. On her daily walks through the farm, she found delight in whatever details caught her attention: a covey of quail, the wheat starting to head out, an interesting cloud formation. She baked cakes and cookies to share with us kids, never once complaining that her diabetes prevented her from tasting them herself. Sister's special talents, though, were drawing and calligraphy, and she created simple, elegant cards for the birthdays and anniversaries of each relative in her large extended family—even for us, her great-nieces and great-nephews. The cards, of plain white stock, bore a personal message in her black calligraphy, and always carried an illustration of a South Dakota wildflower on the front.

The flowers in front of me now in this mountain meadow shine with so much more color, appear in such greater abundance, than the flowers Sister drew. Even this exuberant creek splashing past me will eventually join river after river—the Wind, then the Big Horn, then the Yellowstone—slowing and quieting with each union, until this icy water joins the lazy Missouri that stretches through the prairie where Sister and I both grew up. Nothing on the prairie has the color, the speed, the thrill of these surroundings. Yet Sister, who had traveled so much, including to these Rockies, believed in the beauty of the prairie above all other places. What could she see that I missed?

It was my move away from the prairies, to a Catholic college out West, that, ironically, drew me farther from my faith. I remember that Sister was thrilled when I received a scholarship to a well-respected college, and she gave me her blessing as I left the plains. For my first years there, I spent my days immersed in a holistic blend of academics, daily Mass, and wonderful friendships. I considered following in Sister's footsteps by becoming a nun myself and even spent a weekend at a convent in Idaho, enamored with a life of service, simplicity, and relative solitude.

However, the larger world, with so many other ways of believing, lured me into a period of searching and exploration, not uncommon for twenty-somethings. Gradually, I stopped attending Mass as often, spending hours reading comparative religion and philosophy texts. I learned of the dark side of the Catholic Church, the side that condoned atrocities during the holy wars of the Middle Ages and turned its eyes away from the bloodshed then going on in Latin America. I fell out of love with the church and in love with the world outside my window—a world that included my future husband and a lifestyle that had little room for churchgoing.

Shawn's cowboy credo, the belief of his family, was that God's presence in the natural world provided communion enough with the divine; no manmade ritual was necessary. The freedom of this form of worship appealed to a prairie girl whose life had been filled with the duties of holy days and Lenten fasts and whose spirit longed for more than a childlike obedience to any religious organization. In Shawn's church, the high mountains were cathedrals, the running streams holy water. You alone were both priest and parishioner; the creatures around you your brothers and sisters. Rocks and logs made the pews, and the flowers . . . the flowers were the stained-glass windows depicting the glories of creation.

So here I sit, attending what's become my Sunday morning worship. My daughters, Laura and Carmen, stay home with their dad; this solitude is what truly refreshes my soul, not wrestling with two toddlers during Mass. I think of Sister, who lived in a community of nuns all those years, attending Mass and prayer times daily without question, and I wonder if I'm simply selfish, wanting to spend my Sundays here instead of taking my children to church as I "should." But there are clues in Sister's life that point to a woman who also adored the natural world: she never missed her daily walks; she spent joyful solitary hours drawing flowers and animals; and she even chose her religious name, Margaret Francis, from the Catholic saint, Francis of Assisi, whose love for birds and animals made him the patron saint of all creatures.

Although Saint Francis is a person I recognize from my catechism, it is the saintliness of Sister Margaret Francis that truly inspires me. Her persistence in spite of disease, her pursuit of excellence in both her teaching and her art, her ever-present smile are all marks of sainthood to me. She wasn't a perfect person, and I know she hurt those who loved her at times, but in the end, I think mostly goodness shone through her life. What is different between Sister and me is that while she valued her solitude and took time to feel the glory of the natural world, she still honored her vow to the church and her community, as well. It must have been hard, leaving her beloved prairie at the end of each summer for the more urban, more regimented world of the convent in eastern Iowa, but she did. As I reflect on Sister's inspiration, I must ask myself the uncomfortable question: Am I living up to her standard?

I hike back down the valley to where I left my car, passing large boulders on my left. They look like ordinary boulders from the road, but anyone who takes the time to investigate further will be fascinated to find large white stick figures drawn on them. Speculation about these petroglyphs tells us that the Shoshone Indians drew them hundreds of years ago. I smile to note that these people, just like Sister, valued their connection with the natural world enough to draw it, to find God in it. But I know enough about tribal culture and history to realize that ritual was also a part of these artists' lives; that even if they spent part of their short lives in creative solitude, they also made time for communal worship, for ritual and reverence.

Early Sunday morning, June 2005. Late-blooming lilacs from my neighbor's bush perfume the backyard as I step out to walk the dogs. The scent instantly carries me back to the prairie, to the lilac windbreak around our Dakota farm. I'm not looking forward to walking the dogs today, although I normally love to hike with them. It's just that our move from the country to town last fall has turned this task into one more chore, and duty can steal the joy from any pleasure. I feel the same about attending Mass later this morning.

But as I set out along the creek, my strides lengthen and my breaths deepen, and I begin to enjoy this time. And then I begin to notice the little things: a quiet rabbit hiding beside sagebrush, the trill of a meadowlark, a solitary wild daisy.

Later, in church, the little things again bring me joy: lifting my voice to sing the familiar hymns, helping my daughters follow the Gospel reading, feeling my husband's hand in mine during the Lord's Prayer. My greatest joy, however, comes in watching Laura and Carmen, no longer toddlers, receive the Eucharist in the sacred ritual that connects them to their family, to the congregation around them, to their ancestors, and to the world. It is much the same as those ancient hunters who used ritual to connect their tribe, as Shawn and his family used the annual elk hunt to remind themselves of the divinity all around them. Some parents believe that we should not force any system of belief upon our children but, rather, allow them to choose their own way upon maturity. However, if children don't ever experience God, will they know to go looking for Him or Her? That was the nagging question, the black doubt, that finally got me back into the pew. Shawn had already been baptized, even before we conceived Laura; he knew, but he needed me to lead our family with him. It has always been the duty of the elders to point the way for children, from ancient tribes to Shawn's hunter family to me. Our two daughters have now been joined by four other children, including a boy, Cody Francis, who was named both for his mountain-loving grandfather and his prairie-loving great-great aunt. My duty to them now is as sacred to me as any set of religious vows could be. But among all the lessons that Sister Margaret Francis taught me, one stands out: duty, if embraced fully, becomes not dreariness but a path to joy.

This afternoon, we're driving up to the mountains to look at wildflowers, pointing out to our children all the grandeur of God's church in the mountains, as well. Tomorrow I'll practice my own art by painting pictures with words, practicing my craft on a desk adorned with pine cones, a statue of Mother Mary, and one of Sister's last hand-drawn cards. And, throughout the busy, duty-ridden days of our lives, I'll continue to look for the bright spots of joy, which sparkle like a flower hidden in the grass of the South Dakota prairie.

EPILOGUE

I sit at my computer, in the makeshift office space I've taken from a corner of the mud room, and lift my fingers to the keys, hoping for the words to flow. After nearly ten years of writing, the process hasn't become any easier, especially on days like this, when distractions surface from every direction. Today my husband and a building crew are knocking a hole through the living-room wall to make a hallway into the new addition. The noise and dust, coupled with three preschoolers who obviously won't take naps, threaten to lay waste to any plans I have of writing.

We're in the process of making a home, again. Last summer, after less than a year of living in town, we accepted a ranch job and moved away from the community where we had lived for six years. Now, along with doing regular ranch work, writing, and raising six kids, Shawn and I spend hours restoring and remodeling this 104-year-old ranch house. This morning I painted part of the trim in Laura's room—spring green against lavender walls—and added more splotches to my worn jeans and paint-stained hair. When the crew isn't here building, Shawn piles up old junk and machinery and waits for snowfall so he can burn it. Cleaning up this place and making it comfortable for our family consumes much of our time and attention.

Some would call us crazy. For one thing, we don't own this house; it comes with the job and will be home only until this job ends, whenever that is. We had an opportunity to buy a home of our own last summer, just before we moved. We chose not to take that road—at least not yet—not to saddle ourselves with mortgages, property taxes, and the like. At our age, we probably should be more concerned about investing in our future, building a secure nest egg for our retirement. But the house was in town and not a good fit, in spite of its being a smart investment in that particular housing market. Instead, I quit my job and Shawn took a decrease in salary, all so we could come back to ranching. We don't have much that would impress our peers: no beautiful home, no fine furnishings, no fancy cars. But we have freedom here—the freedom to spend time with our kids and animals, to do work that is meaningful, to enjoy nature and the life we've been given.

And still, despite choosing not to own a house, we seek a home. That search—for a physical and spiritual home—has been at the center of my life. It's a quest that most of the women I know have embarked upon at one time or another. German-Russian immigrant women planted lilac bushes and cooked traditional foods to make the strange, open prairie feel welcoming. My Grandma Quinn, after arriving in South Dakota from Iowa, grew flower beds and baked cakes to establish sensory memories of home. My friend Linda transplants South Dakota wildflowers from her ranch into the lawn at her home in the city to evoke a sense of place. Faced with new beginnings, new places, women rely on pieces of the past to establish home.

In that sense, I am lucky. We have transplanted our family again, but the eight of us are still together. We have all our cats, dogs, and horses, including Sis and Milly, here with us. We have my piano, Shawn's saddles, the kids' toys. I have treasures, like the letter my dad wrote me, the afghan Grandma Quinn knitted for my wedding gift, one of Sister Margaret Francis's last hand-drawn cards. I have the important pieces of our past still with us. So many women in this world have less. This summer, as we peeled layers of old wallpaper from Maria's bedroom walls, a tornado devastated a town just one hundred miles away. Late in August, as we drove to the ranch with the final load of belongings, residents of New Orleans were being warned about the approach of Hurricane Katrina. In October, as we were painting the mud room, Rita hit the Gulf Coast region. As we dealt with new school adjustments and new job frustrations, thousands of people in the Middle East, Africa, and elsewhere, lost homes, families . . . everything. Women everywhere have had to start over in situations much worse than anything I've ever handled.

So I know that much of the depression I faced in late fall and winter was, in the great picture of life, unjustified. And yet I battled through long, dark days when I didn't care if the walls got painted or the meals got cooked. I spent hours aimlessly surfing the Internet, looking for some gratification, some connection to the larger world. I let Cody, Emily, and Katie play alone upstairs, opening the way for all kinds of mischief on their part, and then lost my temper when they did misbehave. I cheered for Laura and Carmen through their basketball games, then came home to pick on them about untidy rooms and undone chores. When Maria needed to read aloud to me for homework every night, I sighed as if listening to her expressive voice was a chore, not a pleasure. A chest cold struck just before Thanksgiving and lasted through Christmas, putting me on the couch and out of life for a good while. Ancient Chinese wisdom holds that respiratory illnesses are often related to grief—an interesting connection if I consider my family's extensive history of these illnesses—and I was grieving.

I had loved Ucross and the Clear Creek valley, and even when we moved into town, I was still able to drive up the valley to check on our mares pastured there. We could still go to the Big Horns for hikes and camping trips with ease. I had a job at the high school, students and colleagues I enjoyed, book clubs and writing groups for camaraderie. My children enjoyed many friends, a great school system, ample opportunities for recreation. Grateful as I was for an opportunity to return to ranching, I needed to grieve all we had left behind there.

Just before this last depression started, I enjoyed a rare opportunity to take a trip alone. Just me. I traveled to South Dakota, to one of the universities there, to give a reading of my work. Along the way, I visited my mom and two of my sisters and enjoyed uninterrupted conversations with them, a pleasure I haven't had much in all these years of having kids. I drove long, sweet hours along highways both new and familiar. I stayed at the apartment my sister Diana uses when she's at the university pursuing her doctorate, and I spent one entire evening with no television, no bedtime rituals, no last-minute homework checks. The highlight of my trip, however, came on Saturday morning, when I picked up Justin, my college-age nephew, and took him to breakfast. I hadn't really spent time alone with Justin since he was a toddler, when I spent weeks each summer babysitting him and his younger brother, Shayne. Simply talking to Justin about his life, his upcoming decisions, and his worries made the entire trip worthwhile, despite lukewarm responses to my reading the day before.

Although the trip will forever be one of my most treasured memories, I know it also spurred the depression that followed. For one thing, my work did not receive the response I had hoped for, so in a financial sense, it had been an impractical investment at a time when we had no money to waste. More than that, though, it led me to question what I was doing with my life. My sister Diana, younger than me, pursues yet another advanced degree; I still hold no more than a bachelor's degree and recently left the field of education—my respectable career—for the second time. My sister Danette holds an important job for the State of South Dakota and owns a beautiful, comfortable home. My brother David, despite legal and financial battles that have left him drained, still owns part of our farm and will fight to keep it in our family. Even my youngest sister, Michele, has established a home business that is more successful than my writing has ever been.

At one point in my conversation with Justin, I found myself giving him advice. He was talking about the degree he pursues, one that is practical and will lead to good jobs but lies in a field he does not necessarily love. I began questioning him about what he truly wants to do, telling him that there must be one thing deep inside that pulls at him—something like my writing does with me. I had been listening throughout the long drive to a CD of scholar and writer Joseph Campbell's *A Hero's Journey* and was really telling Justin to "follow your bliss," an idea of Campbell's that I love. And then I felt foolish; Justin is old enough to see that of all five of my parents' kids, I am something of the joke, the loser. I have six children, but no financial security. I own many horses but still struggle to sell their foals and make a profit. I have spent countless hours these last years writing essays that are largely unpublished; had I put that time into studying or starting a home business, I would be more successful today.

And thus the depression. It finally began to lift on New Year's Eve day, when Shawn and the kids talked me into going out to play in the snow. Guilt is a powerful motivator, and I knew I'd mostly wasted my daughters' Christmas vacation, plopped on the couch with a stack of meaningless magazines. That day, Shawn was taking the kids "four-wheeler sledding"—a flatland diversion to make up for the absence of good sledding hills around here. He ties the sled to the back of the four-wheeler, then drives across the hay meadows, making quick turns and going through whatever dips he can find. The kids love the speed, the sharp angles, the challenge of staying on the sled. The older girls teased me that I could not stay mounted by myself, so I sat down and held on. Shawn drove through cuts and across bumpy frozen cow pies, turning sharply and making every effort to unseat me. And I stayed on. The thrill was in the ride, in feeling my body respond to every change in direction, in staying strong enough to stay on board.

That night, we built a bonfire in the yard. On the advice of one of my favorite authors, Sara Ban Breathnach, the older girls, Shawn, and I had written our regrets and sadnesses of the past year on slips of paper. One by one, we let them fall into the bonfire, watching them burn to nothing. On my paper were all my regrets of leaving Buffalo, of quitting my teaching job, of letting our opportunity to buy a house slip away. I let the paper fall, watched it burn, and then carried my youngest daughter, Katie, into the old ranch house to put her to bed. Katie is named for a character in an old cowboy song titled "Navajo Rug" which tells of a Katie who carries a special blue-and-red Navajo rug through flames as she escapes from a burning building. As I entered our home, surrounded by all our familiar things, I knew I was blessed; I knew I had to move forward into this new year, this new home, this new ride, wherever it might lead.

When I was on my autumn trip back to South Dakota, during my stay at my big sister Danette's home, she and I talked of moving. Years ago, at the same time that Shawn and I had relocated with our three older girls from Dubois to Buffalo to embark on the new job with Bob, Danette and her family had moved from the Black Hills to central South Dakota. At the time, we laughed about the fact that just as I was moving closer, she was moving away. What I didn't know was that she had grieved as much upon leaving her home in the Black Hills as I was now grieving about leaving Buffalo. But in a voice that reminded me so much of my practical German aunts and grandmas, of women who had lived through so much more hardship than I had, Danette talked of how we simply have to make a life wherever we are, have to keep on with the work of raising our children and making homes for our families.

Today, the preface for this book was returned to me by a professor who had heard me read it aloud this fall and wanted to consider it for publication in the literary journal he edited. However, in his rejection letter he told me that the piece, while offering a strong conflict, lacked a satisfying resolution. I think about all the lives I've watched end these past few years without a "satisfying resolution." *Is* there one?

And so here I sit at this computer, trying to tie up in a neat bundle all of the questions I've raised throughout this book, particularly in the preface. I still don't know where I will be buried. I think of that old mother bear often, still wondering whether she knew she was wandering away from the mountains to die on the prairie. Could she have made the decision consciously? Or did she, as I've asked before, just start following the creeks and end up on that particular ranch, in that particular draw? At one time in my life, I wanted so much to believe that she had instinctively known she had to go to the prairie to find her final rest. Now, I don't know that it matters so much. She followed where her path took her, and when her life was over, she lay down in a draw and went back to the earth.

I've lived in many places over the last years. In each, I've sought to establish a home. In some, I've never felt comfortable enough to stay, while in others, I could easily have lived out whatever years I have remaining. As I restore and rebuild this old ranch house, I don't know how I'll feel about this place. But I continue to work to make it ours, just as I continue to write and help Shawn raise horses and step out in faith on the paths in front of me.

One of the saddest pieces of literature I've read over the years is an essay called "Hollyhocks." In it, the author, like me the wife of a ranch hand, has to pack up her life unexpectedly when her husband quits his job. As she is packing, she comes across a packet of hollyhock seeds. Hollyhocks take some time to grow; they are not annuals, flowers that can be both planted and enjoyed in the same season. You have to stay in one place for a while to wait for hollyhocks to bloom. As she prepares to leave yet another home, the author wonders whether she'll ever be rooted enough in one place to feel comfortable planting those seeds.

For me, the initial sadness of the piece was that I, too, have left packets of seeds unplanted over the years, always waiting for the ideal place. But now, the true loss seems to be that the essay's author simply never planted the seeds; never enjoyed the hope that comes with putting seeds in the ground; never had the faith that she would be there to see them blossom. The ideal place will never come, even with thirty-year mortgages and wise investments. How many women in hurricane-ravaged New Orleans thought that the homes they lived in would be theirs forever?

During our renovation of this place, one of the projects has been to redo a chunk of torn-up sidewalk that leads from the muddy driveway to our back door. On the day the cement was poured, just as it was setting up, Shawn and I lined up our children at the end of one slab. One by one, we helped them set their handprints into the still-wet cement: Laura, Carmen, Maria, Cody, Emily, and Katie. Then we wrote, "Acord kids, Dec. 2005" below. Later, as Shawn and I were smoothing out the paw prints that one of the cats had tracked across the slab, I added two small fingerprints in the shape of angel wings, for the baby we lost. It may be the only marker he or she ever has. This spring, I will plant seven tiny lilac bushes along the fence southeast of our house, one for each child, and I will hope to be here to see them bloom.

And we will make this our home, full of our laughter and tears, loving and fighting. I don't know how long it will be ours. I don't know where I will die, or where I will be buried. I hope it will be in a place that has been home for many, many years, a place where my children will be able to sit beside my grave and see places that have been important to our family. But I do know, from the women who have guided me along this path so far, that in the end, what matters is not where I die. What matters is how I live.